IMAGES
of America

THE MORRIS CANAL

ACROSS NEW JERSEY BY WATER AND RAIL

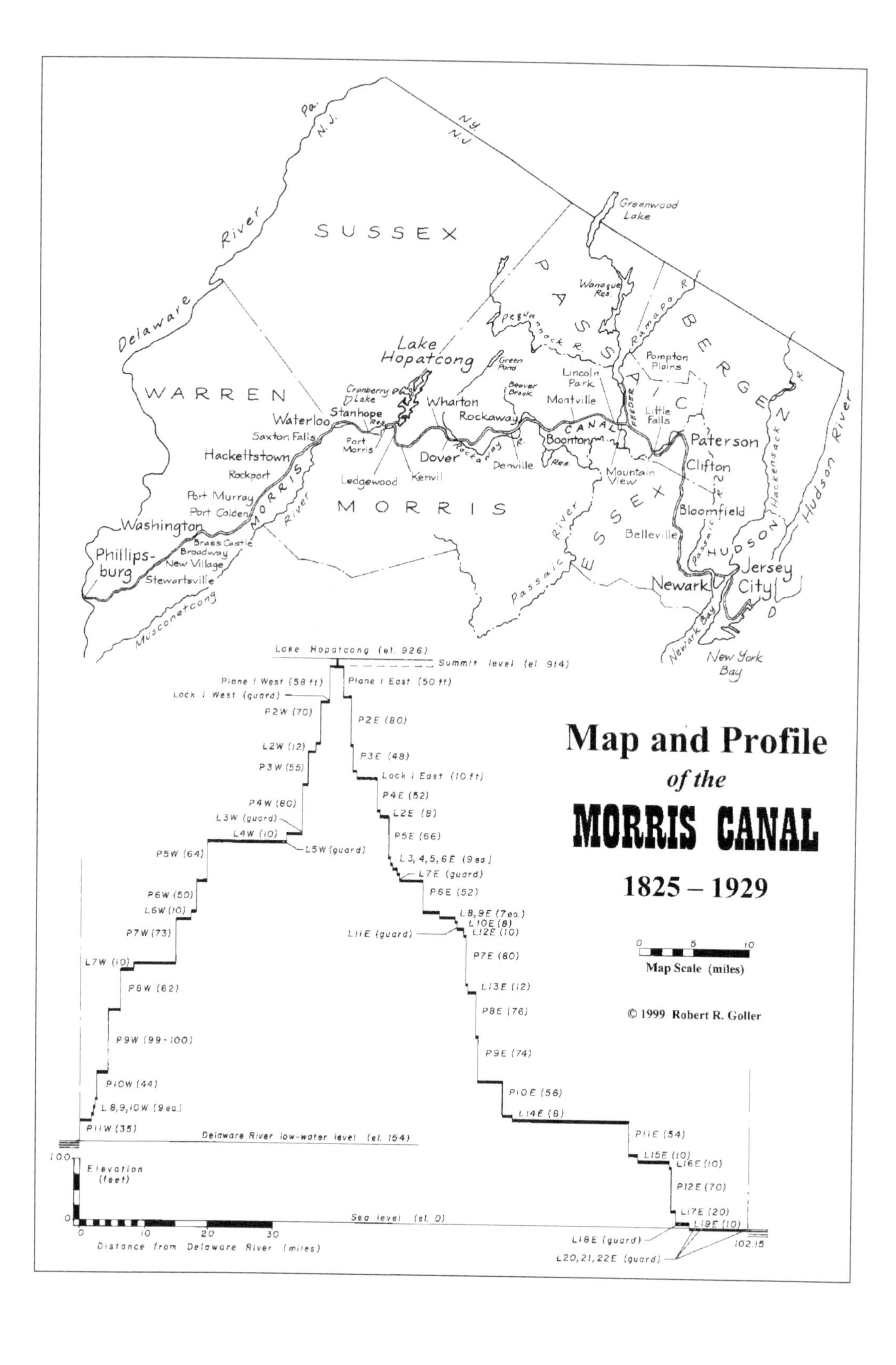
Map and Profile
of the
MORRIS CANAL
1825 – 1929
© 1999 Robert R. Goller
Map Scale (miles)
SUSSEX
WARREN
MORRIS
PASSAIC
BERGEN
ESSEX
HUDSON
Delaware River
Lake Hopatcong
Phillipsburg
Washington
Hackettstown
Stanhope
Waterloo
Dover
Wharton
Rockaway
Denville
Boonton
Montville
Lincoln Park
Paterson
Clifton
Bloomfield
Belleville
Newark
Jersey City
New York Bay
Hudson River
Lake Hopatcong (el. 926)
Summit level (el. 914)
Delaware River low-water level (el. 154)
Sea level (el. 0)
Elevation (feet)
Distance from Delaware River (miles)
102.15

IMAGES
of America

THE MORRIS CANAL

ACROSS NEW JERSEY BY WATER AND RAIL

Robert R. Goller

ISBN 978-1-4671-0410-4

Published by Arcadia Publishing
Charleston, South Carolina

Library of Congress Catalog Card Number: 2019937704

For all general information contact Arcadia Publishing at:
Telephone 843-853-2070
Fax 843-853-0044
E-mail sales@arcadiapublishing.com
For customer service and orders:
Toll-Free 1-888-313-2665

Visit us on the Internet at www.arcadiapublishing.com

Dedicated to the memory of

CLAYTON F. SMITH
1908–1984

founder of the
Canal Society of New Jersey,
whom I was privileged to call
a good friend.

Contents

ACKNOWLEDGMENTS

Many of the images in this book came from the archives of the Canal Society of New Jersey and are so attributed. Within these archives are several collections from which pictures for this book were chosen—particularly the John H. Cunningham–Alice Monahan Collection, the Alfred E. Hagedorn lantern-slide collection, and the Barbara N. Kalata Collection.

The New Jersey State Archives, Department of State, located in Trenton, New Jersey, has an album of photographs and miscellaneous other images made by investigating committees in the early 1900s, a number of which are reproduced in this book.

The Special Collections and University Archives, Alexander Library, Rutgers University Libraries, New Brunswick, New Jersey, is the source for the Harry F. Swayze–John P. Carey Collection, from which all the images attributed to Rutgers University were copied in the 1970s. Donald A. Sinclair and Ronald Becker were my contacts there then.

The New Jersey Historical Society in Newark has the set of lantern slides apparently donated by Dr. H.H. Wilson, which probably should be attributed to Julian R. Tinkham, of Montclair, who may or may not have taken the original photographs; they were used in slide presentations by Tinkham and possibly others to popularize Tinkham's plan for a Morris Canal Water Parkway across most of the state. Some of these are reproduced in this book and attributed to the NJHS; others, not in the lantern-slide collection, were copied from one of the several booklets that were issued by the Morris Canal Water Parkway Association around 1912.

Copies of a few other pictures were obtained in the 1970s from the Newark Public Library, Newark, New Jersey, for which I had the help of Charles E. Cummings. The new chapter in this edition includes important photographs provided by James S. Lee Jr., James S. Lee III, Joseph J. Macasek, Richard Rockwell, and Dr. Richard F. Veit Jr.

Many other pictures are not specifically attributed. Some of these are my own photographs, but I also would like to acknowledge the people who lent, gave, or sold me images that are reproduced in this book. These include William F. Cone, whose photographic work is also held by the New Jersey Historical Society and the New Jersey State Archives; George Coulthard; Olive Davenport; J. Henry Ditmars, whose collection now resides with the Canal Society of New Jersey; Edward Duda; Virginia Faulkner and R. Chris Wolff of Morristown's Old Book Shop; Joseph J. Felcone; Mary Harding; William A. Hubschmitt; John Huntley Jr.; Robert Illig; Carl Maier Jr.; John C. Manna; William J. McKelvey Jr., who obtained the loan of photographs taken in the 19th century by the Orange Camera Club; Harriet Meeker; Harold R. Nestler; Sadie Orts; Margaret Riggin, who lent me photographs from her own collection and from the Historical Society of Bloomfield, New Jersey; Clayton F. Smith; Alma Timbrell; Roy D. Tolliver; Henry B. Whitbeck; and Jesse Wilson.

I am also grateful to Jami Peele, Special Collections librarian at the Olin Library, Kenyon College, Gambier, Ohio, for the portrait of David Bates Douglass; David Breslauer and Lisa Roush of the Macculloch Hall Historical Museum, Morristown, New Jersey, for the portrait of George Perrott Macculloch; and to Nicole Wells, of the New-York Historical Society, New York City, for the portrait of William Bayard.

I also want to thank Robert H. Barth, Richard D. Cramond, Philip E. Jaeger, James S. Lee Jr., James S. Lee III, Joseph J. Macasek, William J. Moss, Richard Rockwell, and my wife, Louise Seney Goller, all of whom helped to make this story as nearly correct as is humanly possible and for whose help and encouragement I will always be grateful.

Introduction

New Jersey's Morris Canal was in some ways a canal of contradictions—a testament to our ability to put a waterway where Mother Nature suggests no waterway should be. As has often been said, it climbed mountains—good-sized hills, anyway—in its circuitous journey across the state. In its final form, it climbed east from Phillipsburg on the Delaware River to its summit level near Lake Hopatcong, nearly a thousand feet above sea level in New Jersey's Central Highlands. From there, it descended in a winding route to tidewater at Newark, on the Passaic River, and to Jersey City, on the Hudson. It meandered about 102 miles across northern New Jersey, over rolling farmland in both eastern and western parts of the state; past great ironworks already old when the canal was built; and through growing industrial towns like Paterson and Newark, then taking their places as part of the greater New York metropolitan area. From the Delaware to the Hudson, the Morris Canal went through a cumulative rise and fall of 1,674 feet—a vertical change believed to be the greatest of any canal ever built.

This unlikely undertaking originated in the 1820s with George Perrott Macculloch. Born in 1775 in Bombay, India, educated at Scotland's University of Edinburgh, Macculloch had by 1810 landed in relatively sedate Morristown, New Jersey, where a decade later he founded the Morris County Agricultural Society. According to his own account, the idea for the Morris Canal came to him during a fishing trip to Lake Hopatcong, probably in the spring of 1822. The basic idea was to tap Hopatcong, the state's largest lake, to supply a new watercourse running west from there to the Delaware and east from there to the Passaic and Hudson. This new transportation route would link the coal lands of northeastern Pennsylvania, the then-extensive iron industry of northern New Jersey, and the metropolitan area of eastern New Jersey and New York City. Macculloch wanted his canal built by the state, but the project was chartered as a private corporation—the Morris Canal & Banking Company—on the last day of 1824. The canal was built from Phillipsburg to Newark between 1825 and 1831 and was extended east from Newark to Jersey City by 1836. In 1837, a 4-mile-long feeder was built from the main canal to Pompton Plains, which brought to the eastern part of the canal an additional supply of water from the Ramapo River and Greenwood Lake.

Macculloch put the problem of carrying a canal across the New Jersey hills in the hands of James Renwick, one of the leading scientific lights of the time and head of the department of Natural Philosophy at Columbia College. His solution was to adopt inclined planes, already used on smaller canals in Europe and Asia, and recently advocated by Robert Fulton as a practical way to overcome large elevation changes on canals. These planes were really small inclined boat railways that carried boats from one elevation to another in special cars riding on rails. When completed, the Morris Canal incorporated 23 inclined planes and 23 conventional lift locks to overcome the 1,674 feet of vertical change from end to end.

The original Morris Canal & Banking Company was also a private bank that could and did issue its own currency. It attracted to its board of directors many men interested in high finance rather than canals, and for a time it made them wealthier than they had been. The Panic of 1837, however, led to a depression that lasted many years, squeezing the life out of many enterprises that had been launched in more vibrant times. The Morris Canal & Banking Company failed in 1841 in the midst of a financial scandal. The bank was out of business from then on, and banking privileges were revoked totally in 1849. The canal itself was leased to private bidders for three years. Late in 1844 a new company was organized, and between 1845 and 1860, as times improved, the canal was entirely rebuilt to larger dimensions under the leadership of capable men. Originally the canal's channel, or "prism," had been 32 feet wide at the water surface, 20 feet wide at the bottom, and 4 feet deep; the enlarged canal was 40 feet wide at the surface, 25 feet wide at the bottom, and generally 5 feet deep. The enlarged canal could accommodate larger, longer boats, but the inclined planes were obstacles to that, since there were limits to how long a boat could be as it maneuvered over a plane's summit or foot and back into the water. The solution was to build boats in two sections that could be kept joined most of the time, but separated where more maneuverability was required; the two sections of a boat were taken across planes independently, in separate cars. Whereas early boats, called "flickers," could handle only about 18 tons of cargo, section boats on the enlarged canal carried up to about 70 tons.

By the 1860s the canal saw a few financially good years. The Civil War pumped vitality into New Jersey's economy, and each year tonnage on the canal grew, most of it coal; it peaked at 889,220 tons in 1866. By then, though, the Morris & Essex Railroad had completed its rail route across New Jersey, closely paralleling the route of the canal. The effect of this competition on the Morris Canal was immediate. Tonnage totals began dropping, and in just a few short years the canal company was already desperate to keep its enterprise operating above a loss. It leased its canal and all appurtenances to the Lehigh Valley Railroad Company in 1871—a perpetual lease of 999 years. This effectively made the railroad the owner of the canal, though the corporate entity known as the Morris Canal & Banking Company continued to exist.

What the railroad wanted out of this was the canal's real estate at Phillipsburg and Jersey City—the valuable terminal properties on the Delaware and the Hudson; in between, the canal itself was mostly a necessary evil. By the 1870s canals in general were no longer viable as transportation routes, and the Morris, now owned by a railroad, was no exception. Its fortunes continued to decline as railroads gobbled up more and more of its traffic, and rumors of abandonment were in the air as early as the late 1880s.

By 1903 traffic on the canal was minuscule. In that year the state appointed a committee of three former governors to determine what should be done with the canal; abandonment seemed to be the obvious answer. Another committee, appointed in 1912, also recommended abandonment, coupled with a program to transform much of the route into a linear water parkway. Nothing came of these studies, though, and the canal, by then virtually devoid of boats, trickled into the 1920s.

What finally killed the canal was a legal case decided in its favor. In 1918 the canal company sued to block construction of the new Wanaque Reservoir in upper Passaic County, on the grounds that the reservoir would divert water needed to keep the canal filled east of the Pompton Feeder. The company won its suit early in 1922, highlighting for everyone that the obsolete canal's legal rights were now impeding the development of a badly needed water supply for this densely populated area.

Before that year was out, the Lehigh Valley Railroad had signed the canal property over to the state of New Jersey—with the notable exception of most of those valuable terminal properties in Phillipsburg and Jersey City. On March 1, 1923, the state took formal possession of the canal and, over the next six years, systematically demolished it.

Even before the end came, the old canal began attracting a new generation of advocates, as people nostalgic for the doomed waterway began to capture its final days in photographs. Along with the photographs taken by the investigating committees in 1903 and 1912–13, it is mostly these pictures—commercially published as picture postcards or individually taken by countless amateurs—that have survived to show us what the Morris Canal looked like. A selection of these pictures, many of them not published before, have been gathered to take you along the Morris Canal of a century ago and more. Enjoy the journey!

—Robert R. Goller
June 1999

One

The Early History of a Technological Marvel

"We spent a delightful day in New Jersey," wrote Frances Trollope (mother of English novelist Anthony Trollope), "in visiting, with a most agreeable party, the inclined planes, which are used instead of locks on the Morris canal." Mrs. Trollope visited New Jersey in 1831; she recorded her impressions at the time as part of her *Domestic Manners of the Americans*, published in 1832. Though her book was generally critical of the American people, she had backhanded praise for those traits that were evident in the building of the Morris Canal: "There is no point in the national character of the Americans," she observed, "which commands so much respect as the boldness and energy with which public works are undertaken and carried through. Nothing stops them if a profitable result can be fairly hoped for. It is this which has made cities spring up amidst the forests with such inconceivable rapidity; and could they once be persuaded that any point of the ocean had a hoard of dollars beneath it, I have not the slightest doubt that in about eighteen months we should see a snug covered railroad leading direct to the spot." The scene above, from her book, probably shows the inclined plane in Bloomfield as it looked then, with a plane car carrying the packet boat *Maria Colden* up the plane. A second pair of tracks (right) carried another plane car, not visible in this view; as one car descended, the other ascended, the whole put into motion by chains attached to the cars and to machinery that was powered by a water wheel running on water from the canal's upper level at the plane. The canal used 23 inclined planes, originally built by local artisans, along with 23 conventional lift locks, to overcome the many changes in elevation across northern New Jersey.

George Perrott Macculloch, of Morristown, supposedly dreamed up the idea for the Morris Canal while he was on a fishing visit at Lake Hopatcong, high in the hills of north-central New Jersey. He saw that natural streams ran westward from there to the Delaware River and eastward from there toward the Passaic; the lake, therefore, could serve as a reservoir for an artificial channel—a canal—that would follow a similar path across the state. He mustered early support for his canal locally by showing that it would be a way to get Morris County's considerable agricultural products to markets in and near New York City—also that it would be a way for Morris County's extensive but depressed iron industry to receive much-needed fuel, and to transport its ore and forge products to iron-manufacturing centers. (Collection of Macculloch Hall Historical Museum.)

The majority of the state's iron mines and forges were in the hills—collectively known as the New Jersey Central Highlands. How would Macculloch's Morris Canal climb the Highlands to reach those mines and forges? Early studies by U.S. Army engineers suggested that more than 200 conventional lift locks of 8 feet each would be required along the whole canal to make this climb. Columbia College's James Renwick (right) suggested an alternative to lift locks espoused not that long before by Robert Fulton—inclined planes.

Renwick's original concept seems to have been to build the planes with two sets of rails, side by side, carrying boats up and down in water-filled caissons, shown in this engraving from a rare Morris Canal & Banking Company banknote of 1825. The caissons would be connected to each other through a system of chains and pulleys. The descending caisson would always have slightly more water, causing it to descend and pull up the ascending caisson. Theoretically, such a system would not have required any external source of power. Renwick's working model of this concept garnered him a silver medal from the Franklin Institute of Philadelphia in 1826.

As built, however, the inclined planes on the Morris Canal used overshot water wheels to move the cars that hauled the boats. These plane cars were open, rather than water-filled caissons. This view, from an 1832 banknote of the Union Bank at Dover, New Jersey, is believed to show the plane built at Boonton in 1829 by John Scott, a canal commissioner who resigned in 1832 to become president of the newly formed Union Bank.

Ephraim Beach, an assistant engineer on the original Erie Canal, was brought from New York in 1823 to survey possible routes for the Morris Canal. Two years later, after the Morris Canal & Banking Company was chartered, he became the company's chief engineer. He oversaw construction of the canal between 1825 and 1836, and went on to build a number of railroads—among them the original stretch of the Morris & Essex Railroad, from Morristown to Newark (1835–1837), and the extension of the Morris & Essex, westward from Morristown to Dover (1846–1847). (Henry B. Whitbeck.)

The earliest planes were built independently by local craftsmen and differed markedly from each other. At best, they were only moderately successful. In 1829 the canal company enlisted David Bates Douglass, an instructor at West Point, to consult on the inclined planes. He became "chief engineer of the planes," which relegated Beach to "chief engineer of the canal." Douglass supervised the erection of those planes that had not yet been built, as well as the alteration of a number of those that already had been built. His stay in New Jersey was short. By 1832 he left the employ of the canal company for other engineering work, later becoming president of Kenyon College, in Ohio, and still later professor of mathematics at Hobart College, Geneva, New York. (Greenslade Special Collections, Olin and Chalmers Libraries, Kenyon College.)

This is one of several types of inclined planes proposed by Robert Fulton, as drawn by him and shown in his 1796 *Treatise on Canal Navigation*, published in England. From contemporary descriptions, it is believed that this design is close to the one used by local millwright Ezekiel Kitchell, of Whippany, who built the first experimental plane on the Morris Canal, at Rockaway, New Jersey, in 1826.

This engraving, from another 1832 banknote of the Union Bank at Dover, New Jersey, is believed to show the Rockaway plane on the Morris Canal as it looked originally.

Some of the early planes incorporated a pair of locks, similar to conventional canal locks, at the top of the incline. These were called "lock planes." This 1836 woodcut shows the plane at the western terminus of the canal at Phillipsburg as it descended northwest to the Delaware River, opposite Easton, Pennsylvania. The drawing may not be a completely accurate rendition, but it is the only known view that shows the locks at the head of one of these early planes.

Other planes ended at the top in a hump, or "brow," that rose above the level of water at the summit of the plane. These were called "summit planes." This view is of the brow of the Boonton plane from a postcard of later times, long after all the planes had been altered to the summit style. Note the plane car on the incline at the left, just over the brow. The incline descends to the left, off the picture.

William Bayard, a prominent New York merchant whose family owned considerable property on the New Jersey side of the Hudson River, was the first president of the Morris Canal & Banking Company. He died in September 1826, however, before much of the canal had been built and while the company was embroiled in the first of a number of financial scandals. (Collection of The New-York Historical Society.)

Bayard's successor was Cadwallader David Colden, a former mayor of New York City and grandson of the last colonial governor of New York. He was elected second president of the Morris Canal & Banking Company early in 1827. Colden restored the public's confidence in the beleaguered company and set it on a proper course. He died early in 1834, after the canal had been completed from Phillipsburg to Newark but before it had been extended to Jersey City.

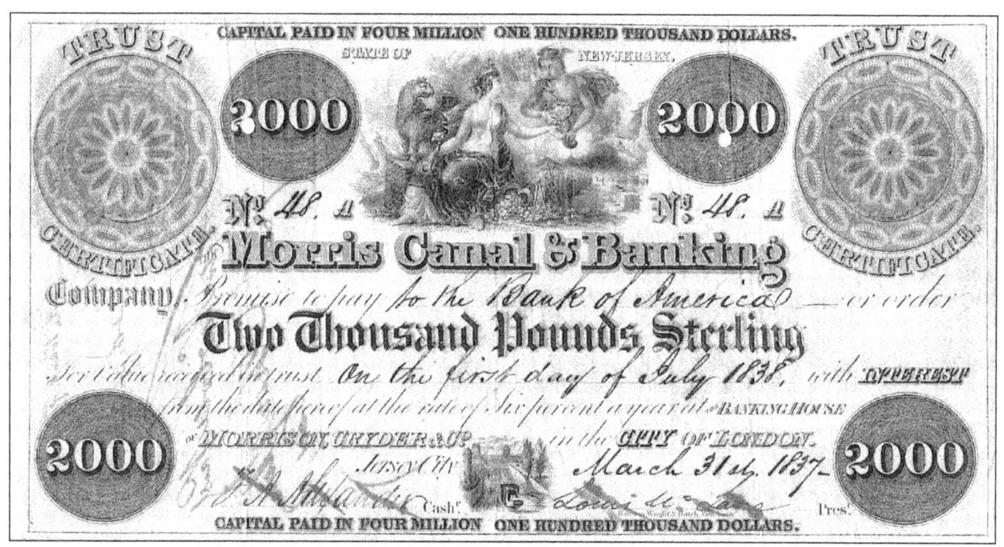

The "banking" side of the Morris Canal & Banking Company was a great attraction to the monied men who served on the company's board of directors. During the 1830s the company's influence extended far beyond New Jersey's borders. The crossed-out signature of the president is that of Louis McLane, who had served as Andrew Jackson's secretary of the treasury and secretary of state. He left the canal-company presidency in 1837 to become president of the Baltimore & Ohio Railroad.

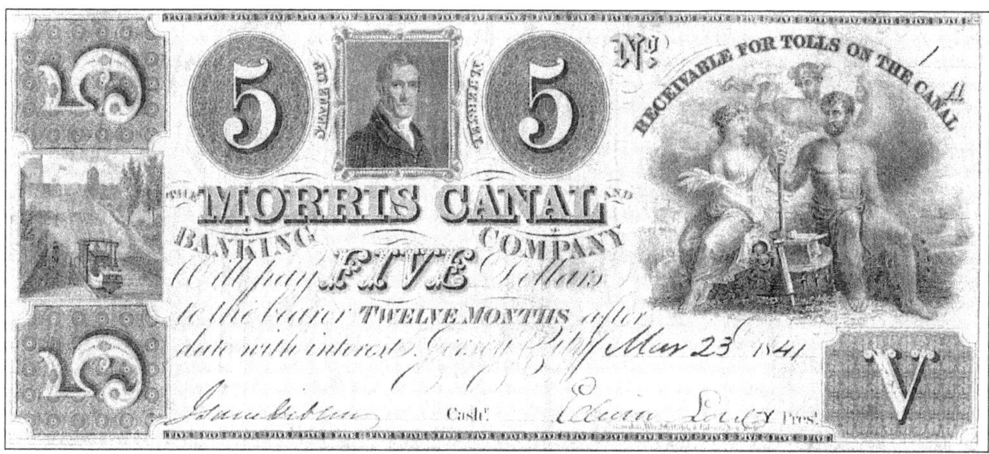

At the end of 1840 the canal company began an ambitious program of enlargement, which it planned to finance in part by issuing "post notes" in payment for work done on this enlargement and in payment of tolls for freight carried on the canal. These notes, issued in denominations of $1, $2, $3, $5, and $10, circulated briefly in 1841 but could not be redeemed at the stipulated "twelve months after date," because the company failed before that. In October 1841 a mortgage on the canal that had been taken out in Holland in 1830 was called in, and the company, having undertaken numerous financial shenanigans, found itself unable to pay.

The canal was leased to private bidders from 1842 through 1844. In the latter year it was sold and reorganized as "The Morris Canal & Banking Company of 1844," whereupon began the second—and more successful—phase of its history. In 1849 it relinquished its banking powers, which had not been used since the failure late in 1841. Between 1845 and 1860 the new company entirely rebuilt its canal—deepening the channel, lengthening the locks, and completely renovating the inclined planes—chiefly under the able supervision of William Hubbard Talcott, an Erie Canal veteran who became chief engineer of the Morris Canal & Banking Company in 1846 and its president in 1864.

A major change Talcott made was in the powering device used at the inclined planes. The overshot water wheels were phased out, each replaced by a "Scotch motor," an early type of turbine, invented by Scotland's James Whitelaw in 1839. These turbines were housed in stone chambers underground, beneath the powerhouses. The Scotch turbine revolved horizontally as water from the upper level of the canal was directed below it, then forced up through it, and expelled through jets at the end of its four curved arms—something like the action of an enormous rotating lawn sprinkler. When the canal was abandoned in the 1920s, the turbine shown above was rescued from Plane 3 East at Ledgewood and was put on permanent display in the state park at Lake Hopatcong, not far away.

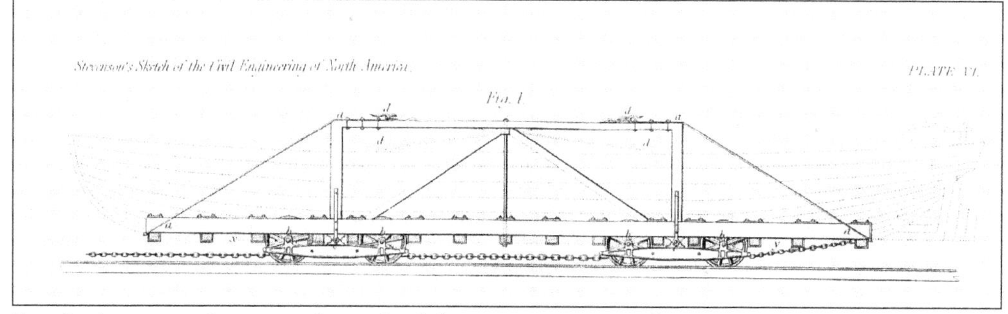

Freight boats on the original canal of the 1830s were small, capable of carrying cargoes of no more than about 18 tons. Because they probably all were gone from the canal before any photographs were taken, we have little information about them. This drawing, from British engineer David Stevenson's 1838 *Sketch of the Civil Engineering of North America*, shows the outline of a Morris Canal boat of that time in a plane car. The car resembles the one in the vignette shown on page 11; no doubt this drawing is an accurate rendition of what was actually in use on the canal.

The boats that came to be used on the enlarged canal may not have been as graceful but could hold much more cargo. These boats, like the one prominently in view at the coal chutes in Phillipsburg, were built in two sections that could be separated for maneuverability at turning basins and inclined planes. The typical section boat of the late 1800s—over 90 feet long with the sections attached—could carry as much as 70 tons of cargo; some boats are known to have carried more than that. These boats were also called hinged boats because of the hinge-pin arrangements used to attach and detach the sections.

Two

PHILLIPSBURG TO HACKETTSTOWN

It may not be unreasonable to begin our pictorial journey even farther west than the terminal of the Morris Canal at Phillipsburg. Pictured above are the coal docks at Mauch Chunk (now Jim Thorpe), Pennsylvania, in the 1880s. Anthracite mined in the hills of northeastern Pennsylvania was loaded at the docks onto Morris Canal section boats like these, which had traveled up the Lehigh Canal from Easton, opposite Phillipsburg, on the Delaware River. Once loaded with coal, these boats returned down the Lehigh Canal and crossed the Delaware into New Jersey. From Phillipsburg, at its western end, the Morris Canal incorporated seven inclined planes and five lift locks on its climb from the Delaware to Hackettstown; from there it rose even higher, over another four planes and through another two lift locks, to reach its summit level near Lake Hopatcong.

This *c.* 1860–63 view westward across the Delaware River from Phillipsburg shows the mouth of the Lehigh River at Easton and South Easton, Pennsylvania. Canal boats from Mauch Chunk locked into the river from the Lehigh Canal just above the Third Street bridge in the distance, continued down the river to the dam, and then turned south, traveling in a channel parallel with the Delaware. Before 1894, those bound for New Jersey then passed through an outlet lock to the Delaware and were guided across by a cable ferry.

In the foreground of this *c.* 1880 view eastward across the Delaware River to Phillipsburg from South Easton is the outlet lock from the Lehigh Canal to the river. A cable (not visible) extended from the outlet lock diagonally across the river to the right, to a tower just left of the small building low on the far bank (to right of center). This cable guided canal boats across the river. The entrance to the Morris Canal is off the picture to the right.

Plane 11 West rose directly from the Delaware at the western end of the Morris Canal. The arch that marks this spot survives today, long after the plane and the canal have mostly faded away. This arch was built below the plane late in the 1860s as a way of protecting the plane from the ravages of river flooding. The arch is open from above, through the center of its masonry, forming a large vertical slot. In canal times this opening accepted squared timbers dropped horizontally from the top, one over the other, to form a barrier in the opening of the arch between the plane and the river. Note the vertical groove running down the visible (right) inside wall of the arch, matched by one on the opposite inside wall; these grooves guided the timbers as they were dropped and helped to hold them in place.

The top of the arch is at the lower left corner in this *c.* 1880 view across the Delaware from Phillipsburg to the dammed mouth of the Lehigh River at Easton. The barrier timbers lie nearby, ready to be pulled out over the arch and dropped into place when necessary. The outlet lock from the Lehigh Canal to the Delaware River and the Morris Canal is off the photograph to the left.

The c. 1880 view above shows the head of Plane 11 West (a 35-foot elevation change) and the incline down to the Delaware River, with the Belvidere Delaware Railroad (left) passing behind the powerhouse for the plane and under the bridge of the Lehigh Valley Railroad over the river. The canal collector's office is at the lower left. An 1894 flood destroyed the outlet lock to the Delaware from the Lehigh Canal, and it was never rebuilt. Thereafter, all coal came into New Jersey by rail—some of it transferred to canal boats here in Port Delaware. As the 1912 view below shows, the unused powerhouse for the plane had been allowed to deteriorate drastically during the years since 1894.

By the 1880s, canal-company operations at the head of Plane 11 West, then called Port Delaware, were extensive. This area included company offices, boat-building facilities, and a large complex of coal chutes, where some of the coal that crossed the Delaware by rail was transferred to canal boats for the trip across New Jersey. This *c.* 1900 view toward the southwest shows the canal property and the Delaware River and Pennsylvania beyond.

On its way eastward, the Morris Canal paralleled the Delaware River along the south side of Phillipsburg, passing a number of industrial sites. The largest was the iron furnace built in 1848 and known originally as the Cooper Furnace; it later became the Andover Furnace. The canal is in the foreground of this *c.* 1905 view, the Delaware River and Pennsylvania in the background.

Near the eastern end of Phillipsburg, the canal passed the area once known as Green's Mills, later as Green's Bridge. The latter name became associated with the large, arched stone viaduct of the Central Railroad of New Jersey, completed in 1865. This view from a later time shows the canal passing under the truss bridge of the Lehigh Valley Railroad, built in 1874–75, and the CRR bridge. Lock 10 West, the first change in elevation above Plane 11 West, is just beyond the bridges, around the bend.

This view is in the opposite direction, from Lock 10 West (a 9-foot elevation change) toward the CRR viaduct. The lock tender's residence is off the picture, to the left. Neither the lock nor the lock tender's house exists today, but the stone viaduct and the abutments of the truss bridge survive.

Locks 9 and 8 West (each a 9-foot elevation change) were close together—almost in sight of each other—but in different townships. Just a short distance outside Phillipsburg, Lock 9 West (above) was in Lopatcong Township; Lock 8 West (below) was in Pohatcong Township. Both of the lock tenders' houses survive as private residences and have been nicely restored, though the locks themselves have been filled in. The building at the extreme left in the view above was the Green's Bridge Hotel; it is no longer there, but the building nearer the lock, which was a canal store, survives as a multi-family residence. Just below Lock 9 West, Lopatcong Creek (above, right) joined the canal for a short distance. The view below shows one of the waste weirs placed along the canal to allow excess water to overflow—in this case to Lopatcong Creek, just to the right.

In July 1912, when photographer William F. Cone took this photograph, the first inclined plane encountered by an eastbound boat was Plane 10 West (a 44-foot elevation change). At this point the canal was back in Lopatcong Township, not far beyond Lock 8 West. The powerhouse is at some distance from the incline, partly hidden by foliage (left of center). This was another place where Lopatcong Creek joined the canal (at center) for a short stretch.

A little more than a mile farther on, eastbound boats approached Plane 9 West. The plane is behind the photographer in this view west from the plane; the canal bridge ahead carries the road for what is today's Warren County Route 519. A pair of plane cars sits in the water at the foot of the plane. The cable that pulled these cars is visible in the extreme foreground, entering the water left of center. (Rutgers University.)

Plane 9 West, shown in this 1903 photograph, was the longest, largest plane on the Morris Canal—one of only three planes that had two sets of rails in the post-enlargement (post-1850s) era. This plane is traditionally said to have had a 100-foot change in elevation, though the elevation change normally may have been closer to 99 feet. The plane tender's residence is left of and behind the three-tiered powerhouse. (New Jersey State Archives.)

East of Plane 9 West the canal approached the outskirts of Stewartsville. This *c.* 1905 view shows Plane 8 West (a 62-foot elevation change) from the railroad bridge that crossed below. The plane cars are carrying an eastbound boat up from the lower level.

At the summit of Plane 8 West, looking back toward the railroad bridge across the lower level, this view shows the "brow" that the plane cars had to surmount to reach the canal's upper level. The little coop on the side of the plane cars was for the brakeman, who traveled with the cars as they ascended and descended. Behind the two men at right is the large wooden flume that carried water from the canal's upper level to the rear of the powerhouse, where it dropped to propel the underground turbine. (Rutgers University.)

The next rise in level was at New Village, which, belying its name, predated the Morris Canal. Lock 7 West (a 10-foot elevation change) was just west of the village. This view by amateur photographer Olin F. Vought shows a canal boat named *The Reindeer* in the lock, headed east on the canal on July 25, 1900. (Canal Society of New Jersey.)

East of New Village the canal passed the equally oddly named hamlets of Broadway and Brass Castle before reaching Washington. This 1903 view shows the stern of a light—that is, empty—boat near Brass Castle headed north ("eastward" on the canal) toward Washington. (New Jersey State Archives.)

At Bowerstown, on the northwest side of Washington, the canal crossed Pohatcong Creek over this stone culvert before ascending Plane 7 West, just off the picture to the right. The culvert remains, now carrying Plane Hill Road over the creek and up the incline, along the former line of the canal.

This *c.* 1905 view shows Plane 7 West (a 73-foot elevation change) from the turning basin below. The brakeman's house (far right) remains as a residence today; the plane tender's house, farther up the plane on the left side (hidden behind trees), was demolished in the 1970s.

The building at left was the original 1850s Morris Canal collector's office at Washington. The smaller building at right was originally an office for the Delaware, Lackawanna & Western Railroad. This 1924 photograph, showing both, is captioned "Offices at Washington," suggesting that the canal company eventually used both buildings. They remain today as private residences. (Canal Society of New Jersey.)

Port Colden, east of Washington, was named for Cadwallader D. Colden, second president of the Morris Canal & Banking Company. On the west side of this hamlet was Lock 6 West (a 10-foot elevation change), shown in a 1912 photograph taken from the head gates of the lock. The lock tender's house, off the picture to the right, is a private residence today.

Beyond the lock was a general store, partly visible at the far left, and a large basin. Port Colden was a busy place for a short time in the mid-19th century, though the boat building and other activities that once were prominent were gone by the 1880s. (New Jersey State Archives.)

William C. Dusenberry, a local entrepreneur, envisioned Port Colden's becoming a major canal port in the 1830s. He commissioned the building of 150 canal boats in 1836, erected an imposing building (above) as a hotel, and advertised building lots for sale. However, nearby Washington became the busy canal port, and Port Colden languished. The hotel became an academy known as St. Matthew's Hall, a hotel again, and later a private residence. It is still there, most recently having been converted for use as office space. Most of the land that Dusenberry offered for sale in the 1830s is still a vacant meadow.

On the east side of Port Colden, straddling both Washington and Mansfield Townships, was Plane 6 West (a 50-foot elevation change). This was the second of the double-tracked planes, though the second set of rails was removed late in the 19th century. The three buildings along the left are canal-company residences. The channel (right) is the tailrace from a sawmill. The plane tender's house is at the upper right, mostly hidden behind the mill building.

The elderly fellow second from left is Abe Scott, for many years late in the 19th century the plane tender at Port Colden. The younger man third from right is William Mayberry, another longtime canal employee, who devised a machine to keep the canal free of weeds. They are standing on the maintenance boat that was fitted on the bow with his weed cutter; the boat is laden with the results of the weed cutter's work.

The terrain at Port Colden required that the powerhouse for Plane 6 West be built some distance from the incline, as at Plane 10 West. This July 1912 photograph by Will Cone shows the powerhouse, left of center; it replaced the one built in the winter of 1847–48, which burned in 1901.

Within easy walking distance of Port Colden was Port Murray, apparently named for James B. Murray, the third president of the Morris Canal & Banking Company. Just before reaching the village, the canal climbed Plane 5 West (a 64-foot elevation change), shown above. At the summit of the plane were John R. Robeson's general store, a blacksmith shop, and a mule barn; these buildings still stand, no longer used for their original purposes, beside the empty canal bed. John Gruver (at left, with daughter Irene) was the plane tender at Port Murray late in the 19th century and probably the last plane tender to live in that plane tender's house. Billy Mayberry succeeded him *c.* 1900 but moved into the brakeman's house, and Samuel Gulick moved into the plane tender's house when he took the job of brakeman on the plane, probably soon afterward.

This clever device, at the rear of the brakeman's house at Plane 5 West, ran the family washing machine from the water of the plane's tailrace. It was allegedly the handiwork of Casper Sutton, the last plane tender at Port Murray.

Beyond Plane 5 West, the canal crossed the main road and passed another canal store (partly hidden by a tree) known in canal times as McCrea's or J.W. Forker's store, and later as Perry's Feed Mill. (Perry's store—which this building sometimes has been called—was a different building up the road to the left.) The old canal store and the McCrea Memorial United Methodist Church next to it still exist. (New Jersey State Archives.)

Another short stretch of canal bed survives in Port Murray east of the McCrea-Forker store. Here, the late poet A.M. Sullivan, who immortalized the Morris Canal in his 1940 epic poem *New Jersey Hills*, makes a new friend along that piece of the old canal on a spring day in 1974. The trees lining the canal would not have existed in canal times, when the towpath had to be kept clear for boat traffic.

On the east side of Port Murray the canal widened to another boat basin, then continued on, curving along the sides of hills on its way eastward. Such hillsides sometimes were the settings for catastrophic washouts if an undermined canal bank gave way.

The countryside east from Port Murray to Hackettstown has looked much the same since canal times. At Rockport the abandoned canal was preserved as part of what became the New Jersey State Game Farm (now the New Jersey Pheasant Farm). This view is from the game farm's early days in the 1930s.

Here is the Morris Canal pay and inspection boat in 1880, when it was the *Katie Kellogg*—near Hackettstown, according to the note on the back of the original photograph. The boat was named for the wife of the superintendent of the canal at that time, William H. Kellogg. It was renamed *Florence*, for the daughter of the next superintendent, William I. Powers, after he rose to that position in 1888. (Rutgers University.)

This southward view (westbound on the canal), shows Buck Hill and the canal on the north side of Hackettstown. The bridge (left) carried the westward extension of Hope Street, Hackettstown—today's U.S. Route 46.

A photographer standing on Hope Street, west of the canal bridge, captured this view in the opposite direction, toward the north from Buck Hill, in July 1914. The spur of water that runs to the left is a canal basin. (Canal Society of New Jersey.)

Three

SAXTON FALLS TO LAKE HOPATCONG

Northeast of Hackettstown, the Morris Canal wound its way into a narrow valley also traversed by the Musconetcong River. At Saxton Falls the canal company had erected a dam in the river when the canal was built, creating a mile-long lake of still water; there, the canal joined the river for a mile of slackwater navigation. Beyond, the canal left Warren County and paid two brief visits to Sussex County—first passing through the village of Waterloo, now a charming restoration open to the public, and the second passing through the larger iron-manufacturing village of Stanhope. From Stanhope it crossed a canal reservoir (now Lake Musconetcong) to Port Morris, climbing Plane 1 West to reach its summit level, near Lake Hopatcong. The 1903 photograph above shows the approach to the guard lock at Saxton Falls, Lock 5 West, where the canal joined the Musconetcong. Because a guard lock is simply a barrier between the canal and a natural body of water, it ordinarily would not have had a lift of a specific number of feet, as at a lift lock; in fact, the change in elevation under normal circumstances should have been close to zero. Since natural bodies of water like the Musconetcong fluctuate in height with natural conditions like floods or droughts, however, there may have been a small lift or drop at any given time. (New Jersey State Archives.)

This photograph, taken from the walkway across the lower gates of the guard lock, shows the head gates and the dammed Musconetcong River beyond. (Canal Society of New Jersey.)

Navigation in the Musconetcong was made possible by this dam, erected in 1830, when the canal was built. The DL&W Railroad passes along on the far side of the river. (Canal Society of New Jersey.)

A mile upstream, the canal parted with the Musconetcong River at Lock 4 West (ordinarily a 10-foot elevation change), visible in the distance at center. Depending on the height of the Musconetcong, the lift at this lock may have been slightly more or less than 10 feet. (New Jersey Historical Society.)

In this closer view of Lock 4 West from the river below, *c.* 1910, Morris Canal boat 796 has just left the lock on its way west. This place was known as Guinea Hollow, and the lock itself was long known as Bird's Lock. Morris Bird was the lock tender here for many years, as had been his older brother Welch and his father, Joseph, before them. (Canal Society of New Jersey.)

The little village of Waterloo, northeast of Guinea Hollow, originally was called Old Andover; contrary to popular belief, the name Waterloo did not appear until about 1840. At Waterloo boats passed through Lock 3 West, another guard lock—this one between the canal and a dammed pond on the Musconetcong known as the Lock Pond. In the *c.* 1905 view above, the guard lock is left of center; from there a bridge crosses the Lock Pond for mules and drivers to follow.

This view at the guard lock is toward the west, away from the Lock Pond. The stone building is the village's general store, still there today. A millrace crossed underneath this lock, making the structure a combined lock and aqueduct. The mechanism that controlled the lower gates of the lock was operated from the platform above the gates.

Right: This young gent posing at the Waterloo lock is believed to be Albert Sherrer. He tended this lock early in the 1900s and later tended Plane 4 West, across the Lock Pond. (Canal Society of New Jersey.) *Below:* This view shows Plane 4 West. Mules and their drivers crossed the Lock Pond to the foot of the plane (an 80-foot elevation change, more or less, depending on the level in the Lock Pond) on the wooden bridge at right. Lock 3 West is to the right of this view. The powerhouse for Plane 4 West, like the one at Port Colden, was set away from the plane tracks; it is visible in the upper right corner. The mound at left is limestone for a plaster mill, which is off the photograph to the left. The bridge in the left foreground crosses a millrace and leads to the heart of the village.

This *c.* 1910 view across the Lock Pond shows the mule bridge (foreground) and a road bridge across the pond (background). The road bridge was destroyed in 1967 by an overloaded dump truck and never rebuilt. (New Jersey Historical Society.)

Continuing eastward from Waterloo, the canal entered Morris County and climbed Plane 4 West. The steeple of the Waterloo Methodist Church rises through the trees in the distance in this view toward the village from the plane. The plane tender's residence is on the left. The bridge that crosses the plane originally carried the Sussex Mine Railroad, which was built to run from the iron furnace at Andover southeast to Waterloo; later, as the Sussex Railroad, it was extended north to Newton and beyond and south to the Morris & Essex Railroad.

Not far east of Waterloo, the canal made another climb at Plane 3 West (a 50-foot elevation change) in the woods of Mount Olive. This 1895 photograph by T.J. Harvey of the Orange Camera Club shows the view down Plane 3 West, back toward Waterloo. The brakeman's house, partly visible down the incline beyond the powerhouse, stands in ruins today in the woods. U.S. Interstate Route 80 now crosses this scene just beyond the ruins of the brakeman's house.

The railroad drawbridge above Plane 3 West carried the Sussex branch of the DL&W Railroad over the canal. Straight ahead along the canal is Lock 2 West. (Canal Society of New Jersey.)

Lock 2 West (a 12-foot elevation change) was tended in the canal's last years by William N. Fluke, his son David, and several of his seven daughters. The stone lock tender's house had a "Dutch oven" built into one wall; the building stands in ruins today, stabilized in recent years to prevent further deterioration. (New Jersey Historical Society.)

The lock tender's house stands off the picture to the right of this view toward Stanhope at Lock 2 West. The canal re-entered Sussex County a short distance beyond this spot.

As the canal approached Stanhope—a settlement locally renowned for being the site of the first anthracite-fueled iron blast furnace in New Jersey, built in 1841—it ascended Plane 2 West (a 70-foot elevation change).

This c. 1905 view shows the incline at Plane 2 West from the upper level. Charles Fluke, a younger brother of William N., was the last plane tender at this site; his residence is hidden behind the trees at left.

The summit of Plane 2 West was just west of Bridge Street (now Kelly Place). From here, the canal widened to a basin. This view shows the bridge and the summit of the plane beyond it from the southeast part of the basin. The large building left of the bridge is John Hulse's general store, which still stands. Recent examination has revealed that the store's walls are concrete, made with slag from the nearby Musconetcong Ironworks furnace.

This c. 1905 view across the basin from Bridge Street, near the Hulse store, shows the rear of Nelden's Pharmacy at left, part of the submerged plane car in the foreground, and the old turnpike bridge over the canal in the distance.

Near the eastern end of the Stanhope canal basin was a plaster mill, built as part of the original ironworks complex, *c.* 1800. In front of the mill was the entrance to a narrow slip that led from the basin to the Musconetcong Ironworks (off this picture to the right). This view, by Jesse Wilson of Dover, is just to the right of the view on the bottom of page 48, though taken at a later time—probably in 1925. The concrete bridge over the entrance to the slip was built as an early part of abandonment work on the canal.

Originally built as the Stanhope Iron Company around 1800, the industrial complex was rebuilt as the Musconetcong Ironworks in the 1860s, after the original company failed. By the early 1900s, when this view appeared on a postcard, it had become the Singer Manufacturing Company.

East of the basin the canal passed under the old Union Turnpike road from Elizabeth to Newton. The *c.* 1914 postcard view toward Stanhope, above, by photographer W.J. Harris, shows the turnpike bridge (now Sussex County Route 183). The view below, in the opposite direction, shows the same bridge, with Lock 1 West beyond.

Lock 1 West, originally built in 1845 as a wooden lock, was a guard lock to the Stanhope Reservoir—today's Lake Musconetcong. Before the reservoir was built the canal continued eastward across swampy lowland to Plane 1 West. (Newark Public Library.)

After 1845 the towpath continued across the reservoir, as seen in this *c.* 1905 view eastward from the Stanhope-Netcong shore. This western part of the route across the reservoir was also the boundary between Sussex County, to the north, and Morris County, to the south. Netcong, in Morris County, is adjacent to Stanhope, in Sussex County, on its south side.

The towpath across the reservoir was apparently a place for family recreation as well as an avenue for canal mules.

On the east side of the reservoir the canal rose across Plane 1 West (a 58-foot elevation change) to its summit level of about 760 feet above the Delaware River—about 914 feet above sea level at its eastern terminus, on the Hudson. (Canal Society of New Jersey.)

Like the powerhouse at Port Colden, this one at Port Morris had been rebuilt after the original burned, in 1886. Its style was noticeably different from that of other powerhouses on the canal. Woodhull Bird, older brother of Morris and Welch, was plane tender here for 60 years—from 1858 to 1918.

Port Morris was more notable as a railroad junction than as a canal town. Across the canal basin at the head of Plane 1 West—in the background in this *c.* 1905 view—was the DL&W Railroad's roundhouse. Many young men who began their working lives on the canal became railroaders.

From the head of Plane 1 West the canal continued eastward through Port Morris. Canal Street, at the left in this *c.* 1914 view by W.J. Harris, coincided with the towpath.

A little farther on, the canal joined with its feeder to Lake Hopatcong. This view shows the bridge for the towpath over the feeder; the main line of the canal is in the foreground, and the feeder to the lake begins under the bridge (left).

In the foreground of this *c.* 1906 view on the 0.67-mile feeder to Lake Hopatcong is "Hypo," photographer W.J. Harris' little terrier, a mainstay of many of his pictures. Under the bridge and around the bend is the lock between the feeder and the lake. "M&E Canal" on this postcard is a misnomer—an example of the common tendency to confuse the names of the Morris Canal and the Morris & Essex Railroad.

The "Black Line" excursion boat *Mystic Shrine* heads for the feeder lock, *c.* 1905. (The lock was 0.67 mile from the feeder's junction with the main canal; this distance appears as "67" on an old list of distances along the canal, which has led to the lock's mistakenly being referred to as "Lock 67.") The Black Line and the White Line were competing transportation companies that carried tourists to the lake from the vicinity of the railroad depot at Landing, near the southern end of the lake. Black Line boats met passengers directly at the DL&W station along the canal; White Line boats met passengers nearby, at the end of a deep channel at the southern end of the lake that brought the White Line's large sidewheelers close to the railroad station.

The lock to the lake, which was unnumbered, was both a guard lock and a lift lock. Though it acted as a barrier between the feeder and the lake, it also made up the considerable difference in elevation between those levels—usually about 12 feet. This lock was also known as Brooklyn Lock, the area having been once known as Brookland, later apparently corrupted to "Brooklyn." George A. Van Wagenen took this photograph while on an Orange Camera Club outing, probably *c.* 1895. Van Wagenen owned Castle Sans Souci, one of Lake Hopatcong's more magnificent early residences.

While the *Mystic Shrine* locks through the feeder lock from the lake, Reuben Messinger, the lock tender (left), walks toward the lower gates. Beyond the upper gates of the lock is a small basin and another single pair of guard gates, which are open behind Messinger in this *c.* 1906 photograph by W.J. Harris.

An early motorboat, the *Dream*, locks through from the lake to the feeder, *c.* 1914, in two views by W.J. Harris. *Above:* The water level in the lock chamber already has gone down a foot or so, as indicated by the wet walls. *Below:* The chamber has almost reached its empty level, and Rube Messinger is about ready to open the gates to the feeder.

The *c.* 1910 view above, taken from high on the southeastern shore of Lake Hopatcong, shows the approach to Bertrand's Island (right) and the western side of the lake (in the distance). The lock tender's house is on the knoll, on the far shore (right of center); the lock itself and the outer gates to the lake are just over the trees of Bertrand's Island (farther right). In the mid-19th century, canal boats were towed across the lake from there to Nolan's Point, farther north on the eastern shore, to take on iron at the ore dock of the Ogden Mine Railroad (below). This traffic ended in the 1880s after the Central Railroad of New Jersey absorbed the Ogden Mine Railroad—just about when the lake began growing as a tourist resort.

Four

Lake Hopatcong to Denville

From its junction with the feeder to Lake Hopatcong, the Morris Canal continued eastward past the DL&W Railroad station at Landing and began its long descent to tidewater. This c. 1905 view shows Hopatcong Station at Landing before it was replaced in 1911 by a new station that still exists but no longer functions as a railroad station. From the station, the canal passed through Dover and Rockaway (towns that had come into being in the 1700s because of the proximity of iron and were revitalized by the canal) and through Wharton and Boonton (towns that came into being and grew because the canal was there to feed that growth). It visited other towns that existed for other reasons—Little Falls, Paterson, and Bloomfield among them. On its eastern descent from the summit the canal passed 12 inclined planes and 16 lift locks that lowered it about 914 feet to the Passaic River at Newark. From there it traveled another 11 miles, crossing the Passaic and the Hackensack rivers into Jersey City and traveling a circuitous route across the Jersey City–Bayonne peninsula before reaching its eastern terminus on the Hudson River.

Just east of Landing was Shippenport, a tiny settlement at Plane 1 East (a 50-foot elevation change). This is a *c.* 1905 view down the plane.

As the canal approached Ledgewood it crossed over the main road that is today U.S. Route 46. The old road passed underneath the canal through this stone culvert, known locally as the "hole in the wall."

Ledgewood—formerly Drakesville—boasted two inclined planes. Canoeists, like this 1912 group, usually had to portage past such places, though some were known to ride the plane cars. These canoeists are pausing at the upper plane, Plane 2 East (an 80-foot elevation change).

This is a *c.* 1905 Harris view of Plane 2 East. The building at the top, just right of and behind the powerhouse, is the plane tender's residence and no longer exists. The dark building (right of center)—the brakeman's house—and the mule barn (center) both survived into the 1970s but also are gone. The white building, a private residence, survives. Much has been done in recent years toward transforming the site of this plane into a historical park.

This is the short stretch of canal between Planes 2 and 3 East, looking toward Plane 3 East, in a *c.* 1906 Harris view. The plane cars and powerhouse for the plane are in the background. The lawn on the slope (right) leads to the Rock Spring House, a local resort.

This Harris view down the incline of Plane 3 East (a 48-foot elevation change), *c.* 1906, shows what is now Emmans Road crossing the plane. Just outside the powerhouse, the large angled beam atop the end of the flume was controlled by the plane tender to open a tub valve, allowing water to drop from the flume to the penstock—a large cylindrical pipe—below it. The rushing water propelled the underground turbine and set the plane's machinery in motion.

The canal continued through the eastern part of Ledgewood straight as an arrow across the Succasunna plain, where construction began in July 1825. This Harris postcard view shows Lock 1 East (a 10-foot elevation change) ahead, and Hypo with one of Harris' assistants along the towpath. The wife of lock tender Benjamin Potter Jackson sent this card in 1907.

In the summer of 1889 the Newark Camera Club chose the Morris Canal for its annual outing and had the use of the *Florence*, the company boat, for the tour. A member of the club poses at the lower gates of Lock 1 East; the *Florence* is in the lock. (Rutgers University.)

From Ledgewood the canal passed eastward through Kenvil. This settlement was originally known as McCainsville, for the McCain family, which operated a hotel, lumber company, and gristmill here.

Beyond Kenvil the canal resumed its descent at Plane 4 East (a 52-foot elevation change). This location was known as Baker's Mills, for the nearby gristmill established and operated by Jeremiah Baker and his descendants. George S. Bird, of the same family whose members tended the Guinea Hollow lock and the Port Morris plane, was the last plane tender at Baker's Mills. He lived in the light building (left) and ran a cider mill in the darker building (right of center).

Amateur photographer Olin Vought captured a boat approaching the summit of Plane 4 East on August 30, 1904. *Above:* The boat is rounding a bend, about to ease into the plane cars; the brakeman already has taken his position on the cars. *Below:* The boat has been floated into the cars, where the two sections are unhinged and fastened to the separate cars. The mules have been unhitched. (Both photographs: Canal Society of New Jersey.)

This Olin Vought photograph of October 10, 1904, has been identified elsewhere as Plane 3 East, but it is Plane 4 East—from this vantage point a fairly similar scene. Water from the flume is overflowing to a channel below that returns to the canal on its lower level, seen in the distance. (Canal Society of New Jersey.)

Olin Vought stood on the deck of this westbound boat to get this picture of the boat being eased into the plane cars at the foot of Plane 4 East on October 24, 1904. (Canal Society of New Jersey.)

Below Plane 4 East the canal entered Wharton, earlier known as Port Oram. At this point it reached Lock 2 East (an 8-foot elevation change). This view shows an empty boat sitting high in the lock, headed west.

Like Lock 4 West, Lock 2 East was known as Bird's Lock. Welch Bird came from Guinea Hollow with his family and belongings on a canal boat early in the 1860s to tend this lock. His son, George S. Bird, tended this lock after him, as did George's brother Stewart and, in 1908, George's son, Charles M. Bird.

The stone lock tender's house had been reinforced long ago with a double "collar" made up of sections of inclined-plane rail across the front and back of both floors, pulled tight by lengths of plane cable. The building burned in 1970 and stands in ruins today; the rail and cable have disappeared. (Canal Society of New Jersey.)

East of the lock, about a quarter mile of the canal remains today, fed by Stephens Brook and looking much as it did in canal times. This October 18, 1897, photograph by Olin Vought shows an empty boat headed west on that section, being towed by a "Jersey team"—a light mule and a dark mule. (Canal Society of New Jersey.)

Wharton was earlier named Port Oram, for Robert F. Oram Sr., a Cornishman who came in 1848 to supervise the building of an iron furnace. He stayed to become a developer of the settlement; this was his store in town, which faced the canal (behind the photographer).

The Port Oram Iron Company was later absorbed by financier Joseph Wharton, of Philadelphia. Reflecting that change, the town changed its name to Wharton in 1902. This is the Wharton Furnace as it appeared *c.* 1905; the canal is off the picture, on the right side of the iron complex.

Plane 5 East (a 66-foot elevation change) was on the east side of Wharton, adjacent to the Wharton Furnace. Olin Vought's photograph of September 27, 1904, shows boats gathering in the basin at the head of the plane. The houses for the plane tender and brakeman are the buildings near center, just before the grove of trees; the top of the powerhouse is left of them. The Wharton Furnace is off the picture to the left. (Canal Society of New Jersey.)

This undated Vought photograph shows Plane 5 East. From this point the canal turned sharply south—to the left in this picture—as it approached Dover, another iron-manufacturing town. (Canal Society of New Jersey.)

South of Plane 5 East, the canal began a succession of small descents. A loaded boat, the *Alert*, approaches Lock 3 East (a 9-foot elevation change) in Olin Vought's September 1908 photograph. (Canal Society of New Jersey.)

After locking through, the *Alert* continues on to Lock 4 East (another 9-foot elevation change) at the bend ahead. The dark building along the towpath (center), beyond the mules, is the residence of the lock tender, who was responsible for both Lock 3 East and Lock 4 East. The Rockaway River is in the valley (left). (Canal Society of New Jersey.)

From the hill on Dover's south side, photographer John Price photographed the canal as it traveled south from Wharton and turned east toward Dover. The stacks in the background are those of the Wharton Furnace. Lock 4 East is right of center.

Olin Vought—who lived in Dover—photographed this eastbound boat approaching the town on August 22, 1905. The millpond for the Ulster Ironworks is visible at the right of the photograph. (Canal Society of New Jersey.)

This bird's-eye photograph of Dover was taken from the hill on Dover's south side. The canal enters town behind the trees (lower left), widens to a small basin that leads to Lock 5 East (center, a 9-foot elevation change), then passes through Locks 6 and 7 East (not visible) at the Rockaway River, bends sharply left, widens again to a basin (slightly left of center), and continues on to the northeast. The large basin (lower left) is the pond on the Rockaway for the ironworks rolling mill. The large church (right, center) is the Hoagland Memorial Presbyterian, which faces West Blackwell Street.

When he was quite young, photographer Jesse Wilson took this photograph (and others) supposedly showing the last boat traveling through Dover. The maintenance scow entered Lock 6 East (a 9-foot elevation change) at Sussex Street; the picture probably dates from c. 1923.

Another of Jesse Wilson's photographs taken at that time shows the guard lock, Lock 7 East, on the east side of the Rockaway River, opposite the Sussex Street lock. The drawbridge for the Central Railroad of New Jersey is being lowered into place after the scow has passed underneath.

Lock 7 East was the site of a bizarre accident on June 12, 1905. With the drawbridge open, a CRRNJ engine lost its brakes and plowed into the bridge, dropping part of the tender on this canal boat while it was locking through. Captain George Meyer, his wife, and their four-year-old daughter were on the boat but were not seriously injured, nor was the engineer, Henry Schafer, of Mauch Chunk, Pennsylvania. (Canal Society of New Jersey.)

Olin Vought photographed these boats on the Dover basin on September 23, 1906. In the background is the Hudson Street bridge. (Canal Society of New Jersey.)

Peters' overall factory was along the canal a short distance east of Hudson Street; this view on a pre-1907 advertising postcard shows the canal in the foreground.

East of Dover the canal turned sharply to the south, then abruptly north at a place known locally as Horseshoe Bend. From Dickerson's Bridge (above), on the bend, the canal continued north into Rockaway.

This 1903 photograph, taken from the cupola of the Rockaway House hotel on Wall Street in Rockaway, shows the canal continuing eastward toward the summit of Plane 6 East. The industrial buildings (right) are those of the Union Foundry. Rockaway, like Stanhope and Dover, was another town built on New Jersey iron long before the canal arrived. (New Jersey State Archives.)

Plane 6 East (a 52-foot elevation change) in Rockaway was the site of the first inclined plane built on the canal, though the structure shown in this *c.* 1910 photograph dates from the enlargement of the 1850s. (New Jersey Historical Society.)

East of Rockaway the canal passed through Denville, in the flat Rockaway river valley. This *c.* 1905 view shows a loaded boat headed east, toward Boonton.

These two views from *c.* 1910–15 show the canal at Peer's store, at the east end of Denville. Lock 8 East (a 7-foot elevation change) was adjacent to the store. Samuel Peer was the lock tender here in 1850, perhaps earlier. One of his sons, Edward C. Peer, probably built the store and took over tending of the lock late in the 1860s. Edward C. was the storekeeper and lock tender for many years, into the 1900s—though the lock tending often was done by others, including two of his sons. The Peer family also had many boatmen on the canal over the years, including one, Roswell B.M. Peer, who was named for canal engineer Roswell B. Mason. Edward C. Peer was also a boatman when he was a young man.

The Newark Camera Club's members pose for a picture at Lock 8 East during their outing on the Morris Canal in early August 1889. The dark-bearded fellow behind the wheel (fourth from right) is William I. Powers, superintendent of the canal; the bearded fellow standing at right, holding one of the upright roof supports, is Daniel "Boney" Osborn, who piloted the *Florence*. (Rutgers University.)

Beyond the lock was a small aqueduct over the Rockaway River. This *c.* 1910 view shows the canal crossing the river into what is now Denville Township.

This side view of the Denville aqueduct was taken from the Denville (west) side of the river. Denville Township is to the right.

This *c.* 1910 photograph in what is now Denville Township shows the canal's eastern course toward Boonton. The distinctive hill in the background is the Tourne, a local landmark and now a Morris County park area. (New Jersey Historical Society.)

Five

BOONTON TO WOODLAND PARK

The Morris Canal crossed the Rockaway River yet again at a place called Powerville, a short distance east of Denville and just west of Boonton. Powerville is today part of Boonton Township. At Powerville the canal encountered another series of locks, of which the first was Lock 9 East (a 7-foot elevation change), shown above in a *c.* 1910 view. The stone ruin (left) was the lock tender's house in earlier days, when Robert P. Farrand tended this lock. One of his many children, William Henry Farrand, who was born in the stone house in 1860, also became a lock tender on the canal, first in Powerville and later at Lock 13 East, on the east side of Boonton. By the time the picture above was taken, James DeHart, who then tended Lock 10 East nearby, also tended this lock. (New Jersey Historical Society.)

Between Locks 9 and 10 East was a small canal basin. Lock 10 East (an 8-foot elevation change) brought the canal to the Rockaway River. This undated view shows Lock 10 East at center and the towpath bridge for mules and drivers at right, as seen from the Boonton side of the river. The lock tender's house, painted red, is nestled in the trees (center). The round center pier in the river for the towpath bridge still exists; the large evergreen tree almost directly over the lock survived into recent times but is now gone. Edward L. Farrand, son of lock tender Bill Farrand and himself a lock tender when he was young, recalled climbing the evergreen in 1910 to see Halley's Comet.

This *c.* 1900 view across the Rockaway toward Boonton shows Lock 11 East, a guard lock, left of center, on the Boonton side of the river. (Rutgers University.)

After negotiating another "Horseshoe Bend," the canal arrived at Lock 12 East (a 10-foot elevation change), near the Pond Bridge over the Rockaway River—a major thoroughfare into Boonton then and now. This c. 1905 view west across the lock shows the canal dredge, or "mud digger," in the canal beyond the lock.

The Rockaway was dammed below the Pond Bridge, and the gatehouse (left) allowed water from the pond to feed into the canal for the use of the canal and the adjacent Boonton Ironworks complex below it, along the river. Lock 12 East is in the background (right). The Pond Bridge is just visible behind the gatehouse (extreme left).

Edward Farrand recalled that this short section of canal between the gatehouse and the summit of Plane 7 East was called the "swift level." The summit of the plane is to the left of this *c.* 1905 view. The ruins of an 1848 blast furnace appear at center right in this picture; other industrial operations were then still going strong.

This *c.* 1910 view down Plane 7 East (an 80-foot elevation change) shows a maintenance scow on the plane cars. Although powerhouses east of the summit were similar, they did vary in some ways. Many were of frame construction, but some, like the one at Rockaway and this one at Boonton, were built of stone.

This *c.* 1900–1910 view shows Plane 7 East (center) with the industrial complex (left) and the town of Boonton (right). The canal and the iron industry were intertwined in a number of ways, not always mutually beneficial. The swift level above the plane supplied both the canal and the ironworks, and it has been said that the water wheels in the ironworks would stop whenever the plane was put into operation.

The Morris Canal took one more step down before leaving Boonton—at Lock 13 East (a 12-foot elevation change). The lock was known variously as Solomon's lock, for Aaron Salmon, who tended it in the mid-19th century; as the Miller lock, for the Miller family that operated a canal store adjacent to the lock; and as the Farrand lock, for Bill Farrand (above), who was the last lock tender here. (Canal Society of New Jersey.)

In Montville, east of Boonton, the canal descended three inclined planes. The first was Plane 8 East (a 76-foot elevation change), shown in Olin Vought's September 21, 1905, view of the loaded boat *Pioneer* headed east. Plane 9 East was just east of this point. (Canal Society of New Jersey.)

This August 1889 Newark Camera Club photograph shows the *Florence* at the summit of Plane 9 East (a 74-foot elevation change). Morris Canal superintendent William I. Powers reclines on the roof of the boat. The building at left, though extensively remodeled in recent years, still stands. The original photograph mistakenly identifies this location as the head of Plane 2 East, in Ledgewood. (Rutgers University.)

This *c.* 1880s or 1890s view of Plane 9 East was taken from the bow of the Orange Camera Club's tour boat. The powerhouse is outlined against the sky (center). The two bridges crossing the plane are both for Main Road (now U.S. Route 202), which doubles back across the incline.

Pontoe's bridge, in the Towaco section of Montville, was one of the more unusual structures on the canal—a cable suspension bridge, shown here *c.* 1910. Company records suggest that this bridge may have been replaced by a more prosaic wooden truss bridge by 1918 or 1919. (New Jersey Historical Society.)

On the east side of Montville, near what is now Lincoln Park, the canal descended again on Plane 10 East (a 56-foot elevation change), shown in this *c.* 1910 view. This area was earlier known as Beavertown, a name that still survives in street names and business names in Lincoln Park. This plane was sometimes also called the Pompton plane.

In Lincoln Park the canal dropped again at Lock 14 East (an 8-foot elevation change). This was long known as Maines' lock; William Maines was the lock tender here for many years. The road bridge shown in this *c.* 1905 view carried the predecessor of today's U.S. Route 202 over the canal. Below this lock, the canal began its longest stretch without an elevation change—the 17-mile level to Bloomfield. (Rutgers University.)

Some distance beyond Lock 14 East, the canal widened to a basin, beyond which it reached the aqueduct over the Pompton River. The view is toward the west, away from the aqueduct and toward the lock. Even in Lincoln Park, this area was known as Mountain View—which, strictly speaking, is across the Pompton River, in Passaic County.

The Mountain View, or Pompton, aqueduct over the Pompton River was built c. 1850. It was 275 feet long and supported on six stone piers, replacing the original 1830 aqueduct, which was 236 feet long and supported on nine stone piers. This photograph was probably taken from the road bridge just south of and parallel to the aqueduct—today's U.S. Route 202.

The building at left in this *c.* 1910 scene in Mountain View was the residence of Edward Hummer, a longtime canal-company employee. He was the last "boss" of the maintenance crew at Mountain View, whose jurisdiction included the feeder north to Pompton Plains. The maintenance scow is moored along the canal bank at left. The eastern end of the Pompton aqueduct is visible at center.

The New York & Greenwood Lake Railroad crossed the canal on this drawbridge in Mountain View. This view eastward along the towpath also shows the towpath bridge over the entrance to the Pompton Feeder (left).

This view shows the towpath bridge over the entrance to the Pompton Feeder from the main line of the canal, which crosses left to right in the foreground. The feeder, built by the canal company in 1836–37, traveled 4 miles north to Pompton Plains, where it joined with the Ramapo River, ultimately bringing a supply of water to the canal from Greenwood Lake and the Ramapo. (Newark Public Library.)

This bridge over the Pompton Feeder carries a pipeline built in the 1890s to bring a supply of water to Newark from the Pequannock River watershed. The Pompton River is in the background. The canal feeder is gone, but the pipeline bridge is still in service.

At the north end of the Pompton Feeder a guard lock joined the feeder to the Ramapo River. This 1905 photograph was taken from the feeder north to the lock and the lock tender's house.

This is the dam across the Ramapo River adjacent to the feeder lock; the lock stands to the right of this picture. From this point, a certain amount of boat traffic continued several miles farther north to the ironworks at Pompton Lake. (Canal Society of New Jersey.)

Farther east along the main line, the canal approached the Passaic River at Totowa and Little Falls. This view in Totowa shows the canal curving past the North Jersey District Water Supply Commission's treatment plant. Water from the upper level of the Passaic, above the falls in the river, is directed under the canal through the four-arched culvert and into the treatment plant (left of center). The canal crossed the Passaic just to the right of this photograph.

The stone aqueduct across the Passaic was built in 1829. Its wooden superstructure was added much later, probably to reinforce the aqueduct, which may have been in some danger of coming apart. (Much of the outside facing already had dropped off by the 1880s.) This *c.* 1920 view is downstream along the river, with Totowa and the treatment-plant buildings visible through the left side of the arch and the township of Little Falls on the right side of the arch.

This is the aqueduct as it appeared *c.* 1860 before the wooden superstructure was added. There was an iron railing on the opposite side, where the towpath was located.

In October 1903 the Passaic watershed suffered the most devastating flood ever recorded for the area. This murky photograph is of that event, taken from a high vantage point, and shows the canal at right, winding through West Paterson (now Woodland Park). The Passaic River and a little tributary, the Peckman River, have converged and are spread across the entire plain at left. The high bridge in the distance is the viaduct of the DL&W Railroad's Boonton branch.

Six

Paterson to Jersey City

At Paterson, the Morris Canal swung around the northern end of Garret Mountain, changing its direction to the south on its way to tidewater. Paterson photographer John P. Doremus captured this view northeast from Garret Rock sometime in the 1870s or 1880s. The canal and the DL&W Railroad (which paralleled it) cut off the small settlement around South Mill Street (below center) from the rest of Paterson; this isolated fragment of a community vanished altogether when U.S. Interstate Route 80 was built on the former canal and railroad beds in the 1970s.

This December 1913 photograph by Will Cone shows the canal's approach to Garret Rock, at the north end of Garret Mountain. From the left edge of this picture, the canal swung sharply back to travel south. (New Jersey Historical Society.)

This *c.* 1905 view is toward the west from about the point at which the canal begins to swing to the south. The city of Paterson is in the valley below (right) and the telegraph poles and stone wall (left to center) define the right-of-way of the DL&W Railroad.

South of Paterson, in what was then Acquackanonk Township, now Clifton, the canal passed through more farmland. In this *c.* 1905 view a loaded boat heads south (eastward on the canal) past the old Centreville Hotel, long known as "Cheap Josie's" (for Joseph Van Winkle, its mid-19th-century proprietor); it was later known as Kesse's Hotel. The hotel was at the intersection of today's Broad Street and Van Houten Avenue.

Farther south the canal entered Essex County, traveling first through the Brookdale section of Bloomfield. This July 1, 1912, view by Will Cone shows the old St. Valentine's Polish Catholic Church and the Franklin Avenue (now Hoover Avenue) bridge. The street at left is East Passaic Avenue, formerly Canal Street.

Will Cone also took this 1912 photograph showing the scene a short distance south of the previous picture—just before the summit of Plane 11 East and the end of the 17-mile level. The building to the far right was a store and tavern at that time; the industrial building beyond it was the Combination Rubber Manufacturing Company.

This wintertime view shows Plane 11 East (a 54-foot elevation change). The Collins homestead (left) was where carpenter Isaac Collins lived in 1830 when he helped build the original inclined plane. The Collins house survives; the plane is now a sloped section of John F. Kennedy Drive.

This view below Plane 11 East shows the James Street change bridge ahead and the small aqueduct over the Third River just beyond it. Change bridges were used where the towpath changed from one side of the canal to the other; when a traveling boat reached a change bridge, mules were led over the bridge, down the other side, and under the bridge to continue on. Not visible in this picture is the Baldwin Street bridge, which crossed the canal just beyond the foot of the plane.

This view is the reverse of the one above—taken from the Third River aqueduct toward the James Street bridge. The plane and the Baldwin Street bridge near the foot of it are mostly obscured behind the right side of the James Street bridge. This and other postcard views mistakenly identify this aqueduct as the Second River aqueduct.

This c. 1909 view shows Lock 15 East (a 10-foot elevation change) and the utility shed at the head gate of the lock. For most of the second half of the 19th century, the lock tenders here were Philip Monaghan and his son Thomas—and possibly also Thomas' younger brother John.

The Second River aqueduct was longer than the one over the Third River. This 1903 view across the aqueduct shows the canal as it bends toward the east, not far from Belleville. (New Jersey State Archives.)

The Morris Canal entered Newark at the northwest end, traveling through the Forest Hill section—in earlier years passing the stone quarries, which were obliterated in the 1890s by architect Charles Olmsted's Branch Brook Park—and originally passing under Orange Street. After a fatal grade-crossing collision between a streetcar and a DL&W train in 1903, the railroad was lowered to pass under Orange Street, and an electric-powered inclined plane was built to carry canal boats *over* Orange Street and back down again. This is a 1913 view of the Orange Street plane. An inverted siphon carried the canal water under the plane, the street, and the railroad. (Canal Society of New Jersey.)

In busy Newark, the canal offered recreation to the city's youngsters, dirty though it may have been. This *c.* 1905 view shows a happy gathering in the Central Ward.

The canal meandered through the industrial city that had been built around it. In this *c.* 1915 view, it approaches a bend that will take it to Lock 16 East. A little of the Gothic architecture of Newark's Central High School, then quite new, shows faintly in the center background. (Canal Society of New Jersey.)

Lock 16 East (a 10-foot elevation change) was the source of Lock Street's name. The lock was located south of Central Avenue, almost directly behind the present campus of the New Jersey Institute of Technology. This view northward from below the lock shows the lock tender's residence (left of center). The last lock tender at this site was William Jewell, who had been a boatman for a time in the 1880s but then turned to lock tending. He continued to live in the lock tender's house after the canal was abandoned. (Newark Public Library.)

From Lock 16 East the canal curved sharply to the east and arrived at the summit of Plane 12 East (a 70-foot elevation change), the easternmost of the canal's inclined planes. This *c.* 1915 view shows the cylindrical flume—unique among the 23 planes on the canal—that led to the rear of the powerhouse. (Canal Society of New Jersey.)

This dramatic view, from the bottom of Plane 12 East, was taken by Will Cone in 1916. This plane was the third with two sets of rails in the post-enlargement era. The tailrace exited between the left pair of rails directly to the lower level of the canal.

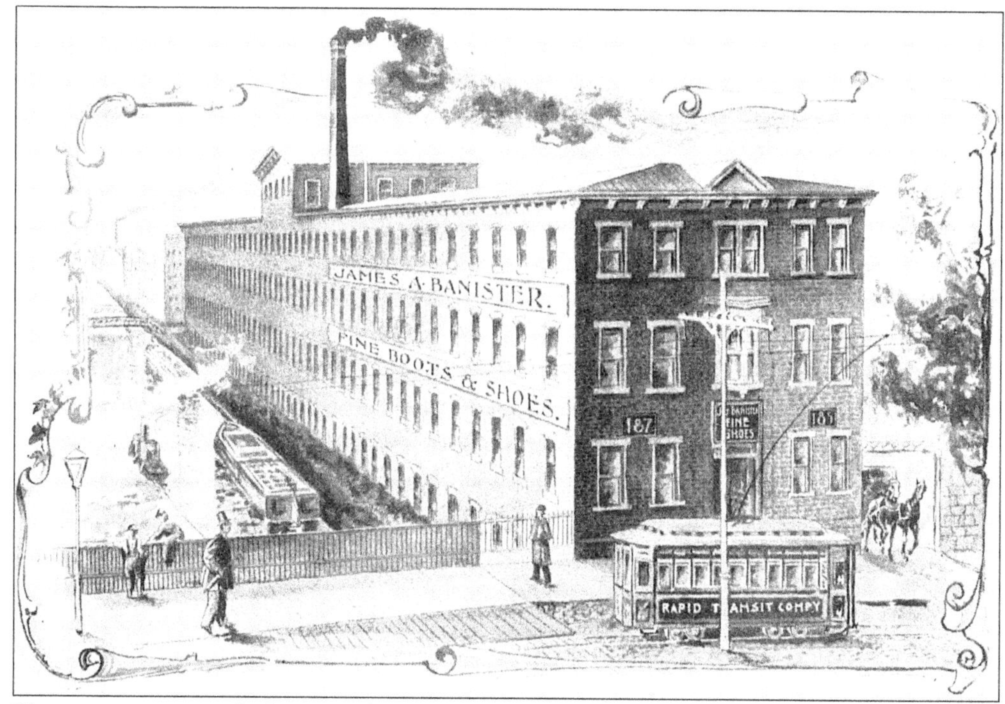

There was no towpath for some distance below Plane 12 East; a steam tugboat (left) towed canal boats to Centre Market, at Broad Street. This rendering from *c.* 1890 shows the view west toward Plane Street from Washington Street. Below Broad Street, two blocks east, boatmen had to push their boats 1,100 feet through the tunnel under the market by poling against the side walls of the canal.

Centre Market was an open area east of Broad Street with a long brick building along its north side (right), under which the canal ran. The Newark Post Office, along Broad Street, is the elegant building with the tower (right of center) in the background. Plane 12 East, not clearly defined, is in line with the market building, to the right of the post office.

This *c.* 1867 view of Newark west from the top of the Fagin & Company flour mill on the Passaic River shows the canal (right) as it emerges from under Centre Market and reaches Lock 17 East (a 20-foot elevation change). The right branch (lower right foreground) is the short spur to the Passaic at Lock 18 East (a 10-foot elevation change, depending on the tide); this was the original terminus of the canal. The left branch, mostly hidden behind the building in the center foreground, is the beginning of the 11-mile extension to Jersey City.

Taken from a street bridge over the Jersey City extension, near the east end of Newark, this photograph shows the view west toward downtown Newark. This is now the route of Raymond Boulevard. (Rutgers University.)

This 1913 view shows Lock 19 East (a 10-foot elevation change, depending on the tide) from below; the bridge carries a branch of the Pennsylvania Railroad. A branch of the New Jersey Turnpike also crosses in this vicinity today. (New Jersey State Archives.)

This wintertime view shows Lock 20 East, the tide lock to the Passaic River at the east side of Newark, with the Passaic River beyond. Canal boats were once towed across the Passaic and the Hackensack, guided by cable ferries; later, after the Plank Road to Jersey City was built, mules crossed the bridges on their own separate walkway. The U.S. Route 1 Truck Route now crosses the rivers at this site. (Canal Society of New Jersey.)

The canal crossed Kearny Point in an open channel when the tide allowed. This Will Cone view west, toward Newark, shows that stretch as it looked on July 1, 1912.

After crossing Kearny Point and the Hackensack River, the canal reached the west side of the Jersey City–Bayonne peninsula. Lock 21 East, another tide lock, separated the canal in Jersey City from the Hackensack. This view is from the river side of the lock looking toward the lock and the basin beyond it. (Canal Society of New Jersey.)

Will Cone's July 1, 1912, photograph west across the pump-house basin at the Hackensack River shows the pump house (left) that brought water into the canal at Jersey City. The water wheel that powered the pump is on the left side of the building. The bridge over the Hackensack is at the right, with Lock 21 East below it, at the far end of the basin.

From the pump-house basin the canal turned south, following Newark Bay down nearly to the Bayonne boundary. This photograph, taken by W.C. Demmert on April 25, 1915, shows Herig's boat club on the bay at the west end of Danforth Avenue, Jersey City. U.S. Route 440 follows the canal's route today. (Canal Society of New Jersey.)

A little farther south the canal turned away from Newark Bay to travel southeast along the Jersey City–Bayonne boundary. This undated view, probably from *c.* 1900, is northwest toward Newark Bay, from this boundary.

Along the Jersey City–Bayonne boundary the canal offered a swath of open space in the midst of what was already a crowded industrial city. By the early 1900s this natural refuge was a favorite spot for people like the ones in the photograph—members of the Greenville Camera Club. The Avenue C bridge is visible in the distance; Bayonne is to the right of the photograph and Jersey City is to the left.

As the canal approached New York Bay on the east side of the peninsula, it changed direction again to travel north along the bay. The bend, known as Fiddler's Elbow, is barely visible at the curve in the distance in this March 23, 1923, view. The photograph was taken from the southernmost of the three Pennsylvania Railroad bridges at Claremont Junction.

By the early years of the 20th century the Lehigh Valley Railroad had taken advantage of its position as perpetual lessee of the Morris Canal by encroaching severely on the canal's right-of-way in Jersey City. This July 1, 1912, view by Will Cone south toward Chapel Avenue shows that the LVRR National Docks branch (right) had elbowed the canal pretty much out of the way.

This view westward, also taken by Will Cone on July 1, 1912, shows the canal aqueduct over Mill Creek just ahead; beyond it are the LVRR drawbridge, the truss bridge of the LVRR National Docks branch, and the plate-girder bridge for Pacific Avenue.

Farther north, the canal made its approach to the Hudson River waterfront, curving from the north to the east, then south, then east again toward the Little Basin, the canal's 1836 terminus on the Hudson River. This 1903 view toward the west shows the line of milk sheds along the LVRR (center) paralleling the canal on its final curve east. (New Jersey State Archives.)

The New York skyline is hidden in the mist in the distance, beyond Lock 22 East, the tide lock at the entrance to the Little Basin on the Hudson, in this March 23, 1923, view east over the lock.

Taken from several stories up, this 1903 view shows the Little Basin and the early skyscrapers of New York City in the distance. Lock 22 East is under the jumble of industrial development in the foreground. (New Jersey State Archives.)

Seven

FADING AWAY

Alvin F. Harlow, author of the 1926 canal classic *Old Towpaths*, took a number of photographs of the Morris Canal in October 1924, while he was preparing his book and just as dismantling work on the abandoned canal was beginning. This scene, which he entitled "The Dead Canal," shows the grassy canal bed in Lincoln Park; the view faces east, toward the Chapel Hill Road bridge and Zeliff's blacksmith shop (left). By the beginning of the 20th century the canal had long since lost out to the railroads, and traffic on the waterway had dwindled. In 1903 the state of New Jersey appointed a committee of three former governors to study the fate of the canal, though all that came of their work was a handsome album of photographs—some of which have been reproduced in this book. The following years saw the brief existence of a Morris Canal Water Parkway Association, dedicated to preserving the canal as a natural greenway for the pleasure of the state's citizens; it, too, failed in its purpose, and a number of the photographs it produced also have been reproduced in this book. Most of the pictures in this book, in fact—whether taken from old picture postcards or amateur or professional photographs—were created long after the Morris Canal operated as a viable water transportation route. By the time photography was available to the general public, which was not much before 1890, the canal was already becoming an object of nostalgia. (Newark Public Library.)

In 1912 the state appointed a new committee to study the Morris Canal problem; this group spent the first four days of July that year and the next exploring the canal from end to end. *Above:* The committee began its 1912 "junket," as photographer Will Cone characterized it, on this scow in Jersey City. Photographer Cone is seated (front center) and Ed Hummer is the steersman (rear). From Newark the committee was transported west to Phillipsburg in the *Mystic Shrine*. (New Jersey State Archives.) *Below:* The committee posed for a picture at Lock 13 East, at Boonton, on July 1, 1912. Chairman of the committee Carlton Godfrey (left) is standing next to committee member Fred G. Stickel Jr., who was also on the 1913 trip and wrote an account of it.

Another decade passed before the Morris Canal gave up the ghost. That happened on November 29, 1922, when the LVRR Company, which had in effect owned the canal by virtue of a permanent lease since 1871, signed the canal and most associated property over to the state of New Jersey. This photograph shows Frank H. Sommer signing the agreement for the state. (New Jersey State Archives.)

The state formally took possession of the old canal on March 1, 1923. Cornelius Clarkson Vermeule Sr., a consulting engineer from an old New Jersey family, who had performed the original topographic mapping of the state between 1879 and 1888, was in charge of the abandonment of the canal. The work took six years, from early 1924 through mid-1929. Vermeule left the project in 1928, and it was completed by sons Cornelius Jr. and Warren Vermeule, along with an associated staff of engineers. The work was officially completed in June 1929.

The state had inherited a corpse. This photograph, one of hundreds taken during the dismantling work, shows the approach to Lock 6 West at Port Colden as it looked on August 13, 1926.

This photograph, taken the same day, shows the overgrown incline of Plane 5 West at Port Murray. The empty powerhouse stands in the background (right).

Alvin Harlow's October 1924 photograph of Plane 10 East, near Lincoln Park, shows the powerhouse and plane cars, the latter already missing one side of their frames. (Newark Public Library.)

In already overcrowded areas like Jersey City, the canal would not be sorely missed. This view of March 28, 1923, shows the canal at Mill Creek, with the LVRR drawbridge, the truss bridge of the LVRR National Docks branch, and the Pacific Avenue bridge—from nearly the same vantage point as the upper photograph on page 111.

While the canal bed sat unused in some places, in other places demolition and reconstruction began almost immediately. At Peer's store, in Denville, Lock 8 East was already being filled in on October 10, 1924, to accommodate a storm-sewer line.

Lock 9 East at Powerville also was disappearing under 20th-century fill; this westward view across the lock was taken October 9, 1924.

Sewer construction also was changing the look of Plane 7 East, in Boonton, as this October 9, 1924 photograph shows.

Not far away, in Montville, the two bridges that crossed Plane 9 East were replaced with earth fill. This is another photograph taken on October 9, 1924.

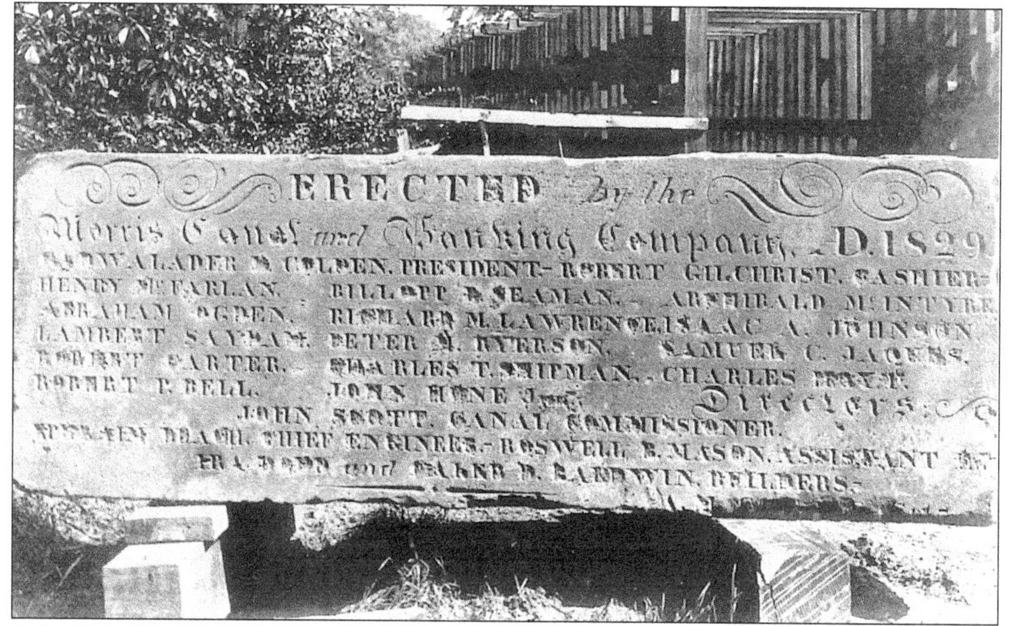

At Little Falls, two sandstone markers, each of which had originally been placed above the arch in the side walls of the stone aqueduct over the Passaic River, were saved for posterity. One had been removed long ago when the facing on the aqueduct began to deteriorate; it had spent the last years of the canal era along the towpath nearby. The other apparently stayed in the side wall until the 1920s, when it, too, was removed. The marker that had been put along the towpath is shown above, up on blocks, with the aqueduct in the background. Eventually both of the stone markers were given a place of honor in Little Falls' Memorial Park, near where the canal had passed. The photograph below shows the other marker as it looked in 1971.

As for the aqueduct itself—the only major landmark to survive for nearly the century that the canal was in existence—it was blasted out of the water in the spring of 1925. Alvin Harlow called its demolition "an act of vandalism, which moved every lover of beauty and history to indignation." The aqueduct did not go willingly; the initial dynamite charge left the central arch standing, and a second was required to totally destroy it. This scene shows the remains of the aqueduct along the Passaic River as seen from the Union Avenue bridge on June 10, 1925, shortly after the event had taken place. A bit of the water treatment plant of the North Jersey District Water Supply Commission is visible in the background (left). In later years the water commission built an underground aqueduct to bring water to highly populated parts of northern New Jersey from the Wanaque Reservoir, built after the canal was abandoned. This aqueduct was to some extent laid in the former bed of the canal and, at Little Falls, its two 74-inch mains now cross the Passaic where the old canal aqueduct had stood since 1829.

"Rusty canallers" like Ed Hummer (above), section foreman at Mountain View, and "Old Bill" Farrand (left), lock tender at Lock 13 East in Boonton, were out of work. Ed stayed on with the canal company, doing small tasks for a while. Bill Farrand continued to live in the old lock tender's house in Boonton. Late in the 1930s he became a local *cause celebre* when he resisted attempts to evict him from the old house; he managed to prevail and passed away in his old home early in 1941. Times were hard for these and other old-timers; not only was their means of livelihood gone, but their final years were spent trying to survive the Great Depression.

The line of the canal itself was transformed in different ways—a subway line in Newark, a short highway in Bloomfield. Above the city of Paterson, the locals were playing bocce in the old canal bed in 1966.

The bocce players soon lost their improvised field. At Paterson the canal and railroad rights-of-way already were slated for U.S. Interstate Route 80, which came soon after the demolition of the canal bed here. This view from 1971 shows the new highway construction from a vantage similar to that of Will Cone's 1913 view on page 96.

In later years some of the old canallers told their stories to succeeding generations. *Above left:* Harry Swayze, who had piloted the *Florence* when he was young, passed on his recollections to Henry Charlton Beck, who wrote extensively about New Jersey's history and folklore. *Above right:* Swayze, at right in this 1950 view, poses with former canal boatman John W. "Ted" Dailey. Many of the pictures Swayze collected over the years later became part of Rutgers University's Special Collections. (Rutgers University.)

In 1946, the canal property at Plane 9 West was purchased by James S. Lee Sr. (right), a Morris Canal historian, collector, and author who has had—as he himself professed—a lifelong love affair with the canal. He and his wife, Mary, the granddaughter of a Morris Canal boatman, raised five children in the old plane tender's house. Looking at artifacts from the plane with him are descendants of Woodhull and Welch Bird. (Rutgers University.)

During the summer of 1971, Jim Lee and a number of neighbors and friends excavated the underground stone chamber that held the Scotch turbine at Plane 9 West, as well as the tailrace that ran from the turbine chamber out to the canal's lower level. This site is now a National Historic Mechanical Engineering Landmark. *Above left:* Jim, at the time of this work, looks down at the chamber from ground level. The chamber itself extends to about 30 feet underground. When the canal was abandoned in the 1920s, most of these chambers were filled in with debris as a safety measure, to keep intruders from falling in and getting hurt. *Above right:* Jim and his crew excavated hundreds of tons of boulders from the turbine chamber, eventually revealing the damaged turbine itself, shown here without its shattered cover plate. *Below:* This is the view up from inside the turbine chamber.

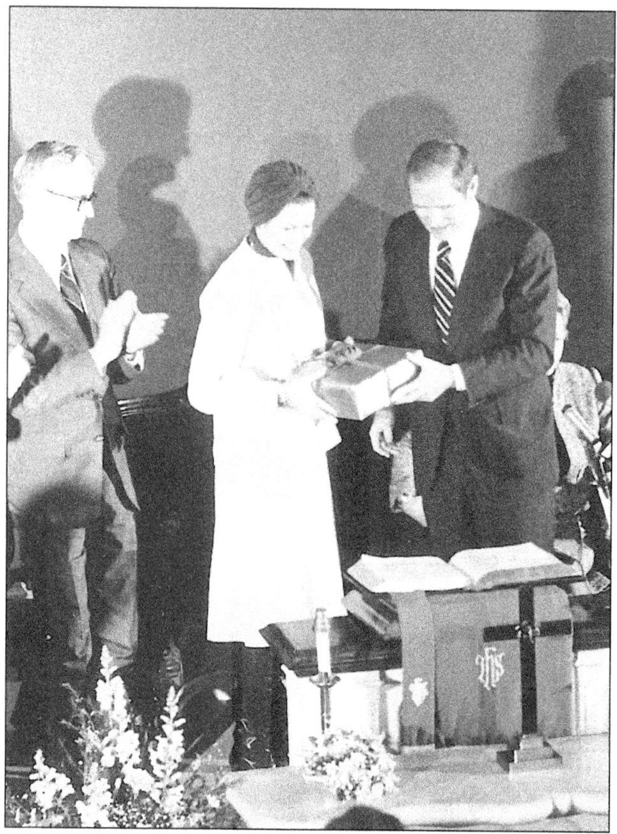

In the 1960s the old settlement at Waterloo was reborn as a restored village, an attraction that has grown in popularity since then. *Above:* In the 1970s a replacement was built for the wooden bridge that had carried mules and drivers across the Lock Pond in canal times. *Left:* Visits by the talented and famous helped put Waterloo in the spotlight—including the visit of former actress Grace Kelly, Princess Grace of Monaco, in 1975 to honor the memory of Irish laborers who helped build the Morris Canal in the 1820s and 1830s. This photograph of the latter event shows Princess Grace, then-governor of New Jersey Brendan T. Byrne, and Dr. Eoin McKiernan, international president of the Irish-American Cultural Institute, at the Waterloo Methodist Church on January 18, 1975.

Right: In 1969 Clayton F. Smith, of Madison, founded the Canal Society of New Jersey, dedicated to the study and preservation of both of New Jersey's towpath canals—the Delaware & Raritan Canal as well as the Morris. Shortly after Princess Grace's visit to Waterloo, the Canal Society of New Jersey opened a small museum in the village to display its collection of artifacts and tell the story of the New Jersey canals. *Below:* Clayton Smith and his wife, Mary Ann, pose with Percival H.E. Leach in front of the museum just before it opened in June 1975. Percy Leach and his partner, Louis J. Gualandi, rescued the mostly abandoned village in the late 1940s and built it into the popular historical restoration that it is today.

Small parks have been established along the line of the Morris Canal with the hope of preserving for posterity shards of what has been called New Jersey's most scenic highway. Bronze tablets and other historical markers have been erected over much of the old route to proclaim for passersby where the old canal once ran. Yet, inevitably, the soul of the canal will continue to fade from memory, replaced at best by a replica of what it once was. All the men and women who plied its waters or worked its locks and planes have now passed on, and the real remains of their residences and the canal itself yield little by little to change or destruction. The time remaining is short; but, for a little while, there still will be, in places, a few hidden ruins that may harbor the ghosts of those long-ago times—for those who discover them, stand in the stillness, and listen.

Eight

The Story Continues

The 20th anniversary of this book is, more significantly, also the 50th anniversary of the Canal Society of New Jersey (CSNJ). The photograph above shows New Jersey historian John T. Cunningham speaking at the inaugural meeting of the CSNJ at Waterloo Village on June 7, 1969. Clayton F. Smith, of Madison, New Jersey, who had been following the history of the Morris Canal for some years, held a meeting at Macculloch Hall in Morristown the previous April to see whether there would be sufficient local interest in New Jersey's two transportation canals, the Morris and the Delaware & Raritan, to warrant forming a historical society that focused on them; he found the attendance overwhelming. There was talk of what to call the organization: Should it be the New Jersey Canal Society, the Canal Society of New Jersey, or something else? Canal Society of New Jersey won out (though people often refer to it as the New Jersey Canal Society anyway). Smith became the CSNJ's first president. He was not the only one who had found the Morris Canal in particular fascinating. Many others, operating on their own—many unaware of the common interest they shared—had been doing so for years. The mystique began long before the Morris Canal was abandoned in the 1920s. Artist and photographer Olin F. Vought, of Dover, began photographing the canal in the 1890s. He created pen-and-ink drawings and watercolor paintings of Morris Canal scenes that he typically gave away to local people he knew. (The CSNJ now owns some of his work.) As a boy in the 1930s, James S. Lee Sr., of Phillipsburg, played among the scattered papers in the canal clerk's office at Port Delaware and rafted in the nearby canal basin and beyond. After returning from overseas duty in the Second World War, he bought what had been the site of Plane 9 West, near Stewartsville. He and his wife, Mary Belle, a daughter and granddaughter of canal people, raised their five children in the plane tender's house there. Lee built an enviable collection of Morris Canal material and had the foresight to interview some of the last living former canal-company employees. Eventually, Lee published what became two popular and well-known books about the canal. Through the efforts of these people and others, the ghost of the Morris Canal gradually acquired a widespread following.

Olin Vought began photographing scenes along the Morris Canal in 1897, when he was 27; over the next decade, he produced well over 100 glass-plate negatives of canal scenes and other scenes of local interest. In 1898, he photographed the pair of section boats shown above, moored along the berm bank of the canal a short distance east of Lock 2 East in Wharton. This photograph became the model for the pen-and-ink drawing below, done five years later. The Roxbury Township Historical Society now owns all of Olin Vought's surviving negatives; the Canal Society of New Jersey owns the original of the drawing. Full-sized copies of the drawing were sold by the CSNJ in the 1980s.

One of the early bus trips of the Canal Society of New Jersey took a group of interested members along the western half of the Morris Canal in November 1969. A highlight of the tour was a stop at Plane 9 West, where James S. Lee Sr. (not yet a published author) spoke to the group about how the inclined plane worked and his plans to see whether the turbine that moved canal boats over the incline was still in place underground. (The weather that day was less than perfect.) Lee is left of center in the photograph above left. The white-haired man is Carl Maier Jr., of Lincoln Park, another who, like Jim Lee, had long been fascinated by the abandoned canal, had collected information about it, painted scenes featuring it, and had given talks about it to local groups. At upper right, the CSNJ group is about to enter the Lees' home, which had been the two-family residence of plane tenders in canal times.

At right, Jim Lee's eldest son, Jim Jr., does the hosting at Plane 9 West on a 1971 CSNJ bus tour. Again Carl Maier Jr. is in the picture, to the left of Jim Jr., and Carl's wife, Alice, is at far left. The grating they are standing on covers the opening to the underground hemispherical stone chamber where, in fact, the great reaction turbine had been found. Its top had been crushed in by an estimated 269 tons of cut stones and debris that had been dumped into the chamber at the time of canal abandonment to keep people from falling in; otherwise, it was pretty much complete.

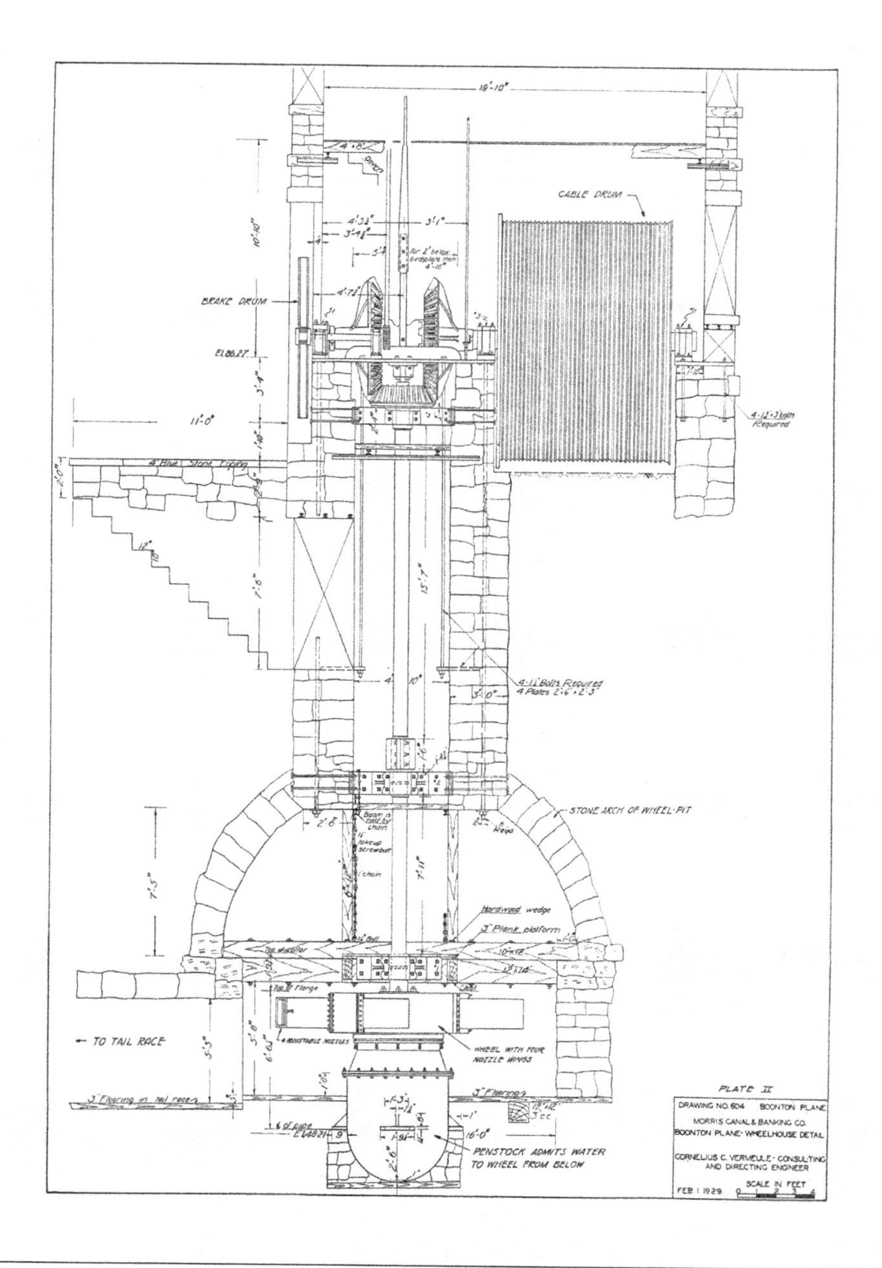

The drawing above was reproduced in the 1929 *Final Report of Consulting and Directing Engineer*, which summarized the work on the state's abandonment of the Morris Canal. It is a cutaway view showing the powerhouse (or in canal-company parlance, the "plane house," also sometimes called the "wheelhouse" or even the "hydro house") and machinery above ground level and below. (The drawing specifically applies to Plane 7 East at Boonton, but the layout is essentially the same at Plane 9 West.) The turbine, penstock, and turbine chamber are shown at the bottom. Photographs that show a little of the exposed turbine and turbine chamber at Plane 9 West can be seen on page 125. The drawings on this page and the next should clarify how the process worked.

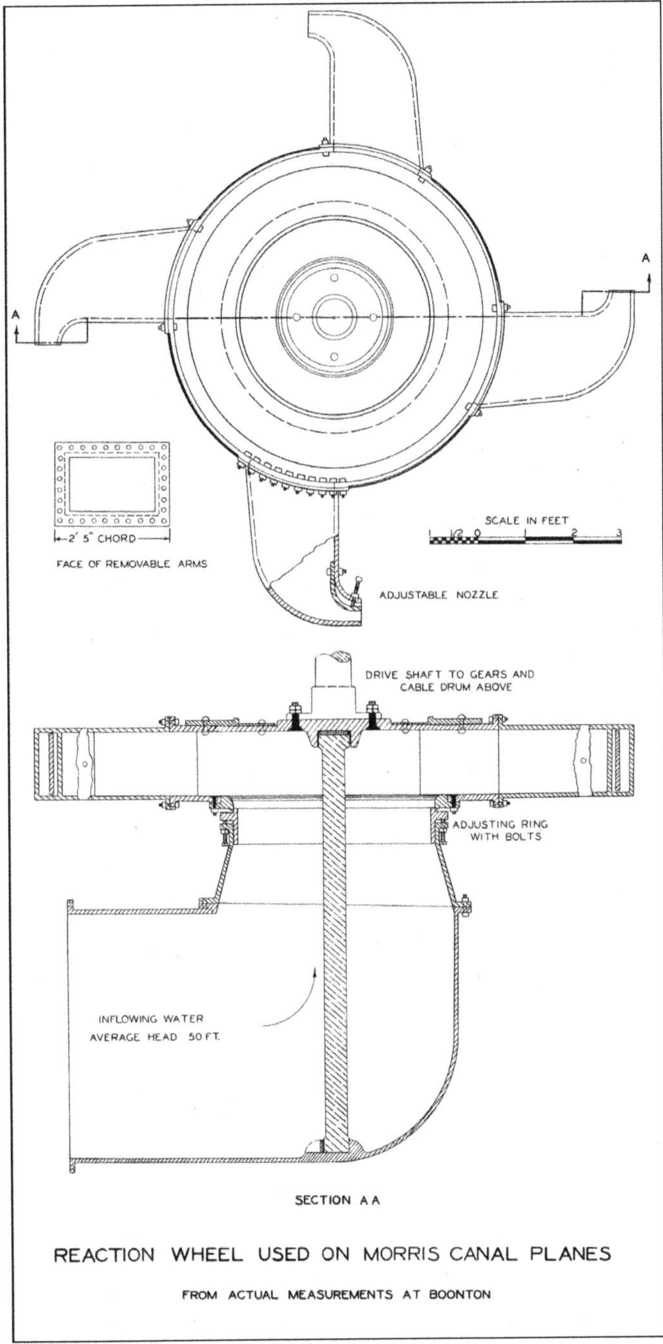

A better representation of what Jim Lee and his sons, friends, and neighbors uncovered is shown in the two-part diagram above (from a preliminary copy of the 1929 final report, not included in the final version). The upper part shows a top view of the reaction turbine at Boonton (again, much the same as the one uncovered at Plane 9 West). The lower part shows the turbine from the side and also shows how the penstock came up from underneath to direct water into the turbine from below, causing the turbine to rotate like a giant old-fashioned lawn sprinkler. The water expelled from the rotating turbine left through the tailrace to the canal's lower level.

A third generation of Lees has also entered the field: James S. Lee III, MA, RPA, a principal investigator and vice president at Hunter Research, Inc., with more than 20 years of experience in cultural resource management, lives in the plane tender's house at Plane 9 West and has participated in a number of Morris Canal archaeological investigations. Jim Lee Sr.'s old office has been turned into a museum that is at times open to the public. The plane tender's house itself is now owned by Warren County and is part of the developing Morris Canal Greenway, an ongoing project that combines canal restoration efforts under a common identity. Above and below are interior views of the museum. (Both photographs: James S. Lee Jr.)

Alice and Carl Maier Jr. (shown above at an exhibit of Carl's canal paintings and other work, including a scale model of a Morris Canal section boat, at the Israel Crane House in Montclair in 1970) also were seduced by the Morris Canal long ago. About a decade before the canal was abandoned, Carl built a summer cabin in Lincoln Park that was expanded over the years and became Carl and Alice's retirement home. Carl and Alice were fond of the outdoors and spent much of their free time canoeing along the canal.

Carl photographed his brother Louis around 1915 standing on the frozen Pompton River, under the Mountain View canal aqueduct, alongside some of the many ice columns that had formed by water leaking from the aqueduct. (Canal Society of New Jersey.)

Another who pursued the history of the Morris Canal on his own was George W. Keppler, born in 1896. He spent much of his adult life working for the Prudential Insurance Company. At left is a photograph of Keppler posing on the stern of a Morris Canal boat in September 1920. Over many years, he sought out what information he could about the canal's history and geography—also meeting some of the old-timers who had worked on the canal. Along the way, he wrote about the canal for Prudential's Newark, New Jersey, employee magazine, *Home Office News*. One of his contributions was a fictional work focused on the *Katie Kellogg,* the canal company's inspection and pay boat. Over several decades, he amassed an extensive archive of material about the canal that remained relatively unknown. (Canal Society of New Jersey.)

Keppler, shown at right doing a handstand on the same canal boat, probably would be forgotten today but for the fact that his niece Barbara N. Kalata idolized him and, many years later, considered it her mission to further his decades-long unpublished research on the Morris Canal. In 1983, with the cooperation of the Morris County Historical Society, she published her exhaustive history of the canal, *A Hundred Years, A Hundred Miles,* dedicated to her uncle. Her collection, which included George W. Keppler's photographs and research papers, is now in the collection of the Canal Society of New Jersey. Kalata also was instrumental in having the Morris Canal officially added to the State and National Registers of Historic Places. (Canal Society of New Jersey.)

In 1964, as New Jersey celebrated its tercentenary, a little book appeared called *The Old Canals of New Jersey*, by Richard F. Veit, who was at the time a member of the faculty of the Westfield, New Jersey, public school system and a part-time member of the faculty in the Department of Geography at Rutgers University. There had been academic studies of the Morris Canal long before, and a book on New Jersey's other transportation canal, the Delaware & Raritan Canal, by Crawford Clark Madeira Jr., had been published in 1941, but Richard Veit's book came at precisely the right time—while interest in New Jersey's history was at a particularly high point. As a child, Richard heard from his father, Joseph P. Veit, "his eyewitness accounts of daily life on the Morris Canal, in Newark, at the turn of the century." Richard's brother Robert, a collector of Jerseyana, contributed to Richard's interest in and knowledge of the old canals of New Jersey. Richard himself was a fellow of the American Geographical Society and a member of the Association of American Geographers, the New Jersey Historical Society, the New Jersey Council for Geographic Education, and the New Jersey Council for Social Studies. Before his book on the New Jersey canals appeared, he had published two other books that dealt with the geography of New Jersey. (Dr. Richard F. Veit Jr.)

The academic is shown at his wall map about the time that his book on the New Jersey canals was published. His son Dr. Richard F. Veit Jr. is now also in the academic world as chair of the Department of History and Anthropology at Monmouth University. He has participated in archaeological work along the former route of the Morris Canal, which probably led to the first of his five books, *Digging New Jersey's Past: Archaeology in the Garden State*, published in 2002. (Dr. Richard F. Veit Jr.)

Some of the people who have been featured here—Olin F. Vought and George W. Keppler—pursued their unusual fascination with the Morris Canal while the canal was still in place (though obviously doomed). Neither one probably ever heard of the other, though their exploratory trips along the old Morris Canal must have overlapped to some degree over time. There probably were others like them whom may never be discovered. Both passed away long before the Canal Society of New Jersey was formed in 1969. The others highlighted on the preceding pages—Jim Lee Sr., Carl Maier Jr., and Richard F. Veit Sr.—also have passed away but were still being lured by the old canal in their own ways in 1969. All joined the CSNJ when it was organized. Many others not included here did the same, and during the past half century, they contributed to the growing knowledge of the Morris Canal's history. Some have referred to this kind of pursuit as searching for something that is not there. To some degree, that is the fascination of it. And every so often, something new turns up, sometimes in unexpected places. Above is a 2016 photograph showing a recently elevated house in the town of Highlands, New Jersey, along the Jersey shore and many miles from the route of the Morris Canal. The house was being raised in anticipation of serious flooding such as was seen and felt along the shore towns during Superstorm Sandy in 2012. (Joseph J. Macasek, Canal Society of New Jersey.)

The contractors working on the raising discovered that the house had been sitting on the front half of a Morris Canal section boat. They removed much of what remained under the raised house and sawed long boards into smaller sections before it was clear that they were dealing with a relic of historical significance. The removed material, however, had been stacked on-site and was still there when industrial archaeologists became involved. The Canal Society of New Jersey was also contacted; president Joseph J. Macasek and longtime member and former director William J. McKelvey added their knowledge of canal boats and the history of the Morris Canal to the joint effort. A detailed report of the discovery and examination of what was found was prepared by industrial archaeologists Jean B. Pelletier and Chris Cartellone of AECOM, Inc., a multinational engineering services company, for the New Jersey Department of Environmental Protection. (Joseph J. Macasek, Canal Society of New Jersey.)

Ultimately the owner of the property agreed to turn the remains of the canal boat over to the CSNJ. The various pieces of the boat were transferred by forklift to a large trailer—work was overseen by Bill McKelvey of the CSNJ and Phil Francis of Shadyside Trucking—and transported from Highlands to historic Waterloo Village, where the CSNJ maintains its museum and has access to a large barn that served in the 19th century as Samuel T. Smith's carriage house. (Joseph J. Macasek, Canal Society of New Jersey.)

From interviews with a longtime Highlands resident who lived nearby, it was learned that this front half of a Morris Canal section boat apparently had been obtained farther north by another Highlands resident, probably along the Jersey City waterfront, and repurposed (with a new post-canal-era rudder) for use as a fishing vessel during the 1920s and 1930s. (It may also have been outfitted with a sail and used as part of a rum-running operation during the Prohibition era, according to the recollections of the interviewed Highland resident's late husband.) For now, the bow and associated parts reclaimed from the Highlands boat reside in the carriage house at Waterloo. A large photograph showing much of the top of a Morris Canal section boat during canal times has been placed on the wall behind the bow. The bow itself and the photograph combine to show, in a rough way, what the top of the two sections of a Morris Canal boat would have looked like. So far, these remains of a fore section constitute all that has been found of the great section boats that once plied the canal. (Joseph J. Macasek, Canal Society of New Jersey.)

Until the discovery of the remains of the fore section of a Morris Canal boat in Highlands, the best that could be done was to create a reproduction. Scale models exist, some of the older ones made by retired canal boatmen themselves, years ago, as remembrances. In more recent times, a full-size reproduction was built at the site of Lock 7 West, at New Village in Warren County, known in canal times as the "Fresh Bread Lock." The area around the site of the lock is now a restored spot called Bread Lock Park. The photograph above shows the stern of the replica boat.

The stone foundation of the lock tender's house also still exists at Lock 7 West in Bread Lock Park. The park is part of the Morris Canal Greenway, the seeds of which developed with people in the Canal Society of New Jersey—Brian Morrell, who had done a detailed study of the Morris Canal in Warren County in the 1980s; Bill Moss; Bob Barth; Joe Macasek (all four of whom have served as presidents of the CSNJ); and many others.

Efforts to memorialize the Morris Canal, now combined into an ongoing project called the Morris Canal Greenway, began before the official abandonment itself was over. In 1925, as work on the new dam at Lake Hopatcong was being completed, the turbine from Plane 3 East at Ledgewood was removed from its underground chamber and set in a concrete shelter at the new park at the lake. It continues to serve as a reminder of the 19th-century technology that allowed the canal to surmount the many elevation changes that were necessary in crossing the northern part of New Jersey. Also during the abandonment years, as can be seen on page 120, the two coping-stone markers that had been in the great stone aqueduct over the Passaic River at Little Falls were saved and eventually relocated a short distance away from the site of the aqueduct in what is now Memorial Park in Little Falls. They are the only major remnants of the original 1825–1845 canal that survive. (The two stones are now better protected from the weather by shelters built over them.) Cornelius C. Vermeule Sr., until 1928 in charge of the official state canal abandonment (his son Cornelius C. Vermeule Jr. succeeded him), made a pitch for saving the entire inclined plane in Boonton (Plane 7 East) as a working exhibit to show people, especially those interested in technology, how the inclined planes operated. As with earlier efforts to save much of the canal itself as a water parkway, nothing came of his suggestion. The planes were all dismantled, some more thoroughly than others. Many of the turbines must have been removed for their scrap-iron value, especially as the threat of a second world war grew—though, as can be seen on pages 124 and 125, James S. Lee Sr. found the turbine at Plane 9 West in place on his property in 1971. His son James S. Lee Jr. lives nearby on the site of Plane 10 West and has seen evidence that the turbine for that plane is still there as well.

In 1973 Donald C. Kuser, a Pequannock Township councilman, spearheaded an effort to create what became Aquatic Park, part of the Passaic County Park System, north of the Pompton Feeder in Pompton Plains. Above is a photograph from a public-awareness event Kuser held that spring at the proposed park area—the maze of small islands north of the site of the feeder lock. Kuser is at center, in the back of the canoe.

In 1976 West Paterson (now Woodland Park), led by Joseph Iozia and a number of local groups, likewise made an effort to restore a section of the canal near Browertown Road as a small memorial to the canal. The photograph above shows a parade along the restored canal bed after a dedication ceremony in the park. These projects at Pompton and Woodland Park, completed long ago, are now part of the growing Morris Canal Greenway.

The photograph above was taken by Newark commercial photographer William F. Cone on December 13, 1913, apparently for possible inclusion in the forthcoming Godfrey Committee report (see page 114). It is one of nine he took that day between Paterson and Bloomfield. His caption for this one is "Morris Canal From North of Allwood Road, Looking South." The road in the foreground is Broad Street, Clifton; the canal bridge in the background is probably for the road that then existed as High Avenue. (The photograph *was* included in the 1914 Godfrey Committee report.) The section of canal bed in the foreground survived along Broad Street long after canal abandonment. In the mid-1980s, Clifton resident and CSNJ vice president Jack W. Kuepfer led an effort in connection with local Boy Scout troops to develop a strip of the area along Broad Street as a small nature preserve and memorial to the Morris Canal. That work, completed in 1987, brought Kuepfer much local recognition and a presidential award as a winner in the national Take Pride in America Program in 1988. The photograph below was taken from roughly the same spot as Will Cone's 1913 view. The park, like the other locations featured, is now part of the Morris Canal Greenway.

Decades after the Morris Canal was mostly wiped off the landscape, archaeologists began to dig up its remains. The scene above shows, third from right, pioneering industrial archaeologist Edward S. Rutsch (one of the founders of the Society for Industrial Archaeology) and his crew digging at the site of the powerhouse for Plane 10 East, at the border between Lincoln Park and Montville, the area known as Towaco, on April 25, 1981.

Above, one of Ed Rutsch's crew clears off the exposed top of the stonework at the foundation of the powerhouse. During the 1980s Montville and neighboring Lincoln Park developed a master plan to create a corridor of open space along the former line of the canal, which included the sites of three inclined planes. This 1981 dig at Plane 10 East was followed in years to come by extensive archaeological studies and reports at Planes 9 and 8 East. In July 2013 Montville dedicated Montville Canal Park, also part of the Morris Canal Greenway, near where the incline of Plane 9 East once crossed Main Road (U.S. Route 202). In the future, the trail from this linear pocket park may be extended to the site of Plane 8 East, not far beyond.

Ground was broken for the Morris Canal, apparently without much ceremony, "near Suckasunny Plains," on July 12, 1825. This was in today's Ledgewood and probably on the land of Silas Riggs, who was one of the canal's original contractors. On July 12, 1975, the Canal Society of New Jersey and the Roxbury Township Historical Society jointly marked the sesquicentennial by unveiling a historical marker comprising two stones—one of them a stone "sleeper" that long ago helped support one of the rails on Plane 3 West, near Waterloo.

U.S. senator Clifford P. Case was one of the invited guests at the July 12, 1975, celebration in Ledgewood. The building behind the elevated stand is the Silas Riggs Saltbox House, headquarters of the Roxbury Township Historical Society. (This historic building had been moved in 1963 from its original location not far away to this one because of modern development.) The building with the corner quoins, partly visible behind the Riggs House, is King's Store, a landmark built in 1815 that later became a canal store. In 1975 it was sitting closed, its inventory frozen in time since owner Theodore F. King died in 1928 and his daughter Emma Louise King closed it. The store, now owned by Roxbury Township, was reopened after Emma Louise King passed away; it has been restored as a museum.

Roxbury Township included within its borders five inclined planes—two of them close together at Ledgewood. The upper plane, Plane 2 East (see page 61), remained in relatively untouched condition, except for the removal of most iron parts, long after the 1920s abandonment. The turbine chamber was mostly filled in, leaving just the vertical walls down to the tops of the arches exposed. The photograph above was taken at the top of the turbine chamber on April 3, 1966. At that time, some of the foundation of the powerhouse still survived in the background. This view appears to be toward the north, to the top of the incline.

Above are single images from two stereo pairs photographed at the top of the turbine chamber at Plane 2 East on November 18, 1978. The image at left is toward the bottom of the chamber from one end; the tops of the arches (below which was the hemispherical chamber that housed the turbine) show at the bottom. At right is a close-up of one side.

With the help of a generous Green Acres grant in 1981, the Roxbury Rotary Club purchased 240 acres of land that included the site of Plane 2 East. The Rotarians cleared the incline; cleared and repaired the turbine chamber, tailrace, and bypass flume; and created Ledgewood's Morris Canal Park, now also part of the Morris Canal Greenway. By 2007, a new effort was made at further restoration, since some of the interior stonework was failing. The photographs above show the tailrace arch at the interior of the turbine chamber before and after repairs were made. (Both photographs: Joseph J. Macasek, Canal Society of New Jersey.)

At the ground surface, the stonework at the opening to the turbine chamber was built up and covered by a grating for safety and a weather-protective pavilion—not exactly a reproduction of the powerhouse, as the Rotary Club originally had in mind, but sufficient to protect the remains from further deterioration. The bottom of the incline is toward the right. (Joseph J. Macasek, Canal Society of New Jersey.)

One of Bloomfield's major canal landmarks was Plane 11 East, seen on page 98 in a canal-era photograph. The incline, stripped of its rails and other reusable metal parts long ago, sat unused long after the canal was abandoned. In the 1950s, when the Garden State Parkway was built, the nearby incline and the canal below it and above it also were paved over and became a bypass road to ease traffic congestion on nearby Broad Street. Originally called the Morris Canal Highway, it was renamed John F. Kennedy Drive after President Kennedy's assassination in late 1963. The photograph above, showing the incline, is from the summer of 1963.

The house to the left of the incline in the picture on page 98 was still there and occupied in 1963; it is lost in the foliage in the top photograph. The earliest part of the house, probably before 1800, was built by carpenter John Collins. His son Isaac, also a carpenter, participated in the construction of Plane 11 East (just behind the house) in 1829–30. Isaac's son John became a master carpenter for the canal company. The Collins House was in danger of being demolished in the early 1980s, but a cultural resource survey by Brian H. Morrell, Herbert J. Githens, and Edward S. Rutsch in 1982 forestalled that. Despite its documented significance, the unoccupied house deteriorated over the following years and only now is in the process of being restored and preserved. The house was added to the State and National Registers of Historic Places in 2017. (Richard Rockwell.)

Another recent Bloomfield project was the replacement of the 1922 Berkeley Avenue bridge—one of the last road bridges built over the canal while it still existed. The 1922 bridge spanned both the Second River and the canal. When the canal was abandoned, the property below the bridge became a recreation trail as part of Wrights Field. The 2009 photograph above shows the portion of the bridge that had once crossed the canal; in this image it crosses part of the recreation trail. (The Second River is at left.) This 1922 plate-girder bridge, with a lattice fence for a walkway, had outlived its predecessor, a truss bridge with a similar lattice-fenced walkway built in or before 1903, by about 20 years. (Richard Rockwell.)

With approval from the New Jersey Historic Sites Council, which required that a new bridge retain the look and character of the existing bridge, the 1922 bridge was replaced in 2016–17 by the one shown above. (The Second River is at right in this view.) This project and the Collins House project are important additions to the Morris Canal Greenway, supported by the town's Morris Canal Greenway Committee, the Canal Society of New Jersey, and other interested parties and organizations. (Richard Rockwell.)

On this page are two charming moments from 1904, captured on Olin Vought's glass plates at the outskirts of Wharton. Above are two young girls walking westward on the towpath, perhaps with goods bought in town, talking with the captain of a passing canal boat, also headed west. For both photographs, Olin Vought stood on the platform over the foot of Lock 2 East—Bird's Lock—at the extreme west end of Wharton.

Above, the girls have stopped to watch a canal boat (company number 762) being lifted about eight feet in Lock 2 East. This photograph carries useful information about how the canal lock was constructed—information that would come in handy more than 100 years later, as we will see.

When the Morris Canal was abandoned in the 1920s, Wharton purchased the right-of-way and retained the quarter mile or so of canal from West Central Avenue to Lock 2 East as a recreation area, later named Hugh Force Park. This section was fed just below the lock by Stephens Brook, a branch of the nearby Rockaway River; it has continued to the present day as a canal prism full of water. For a time, the part closest to West Central Avenue was used as a swimming pool; the remainder always has been used by local fishermen or enjoyed by people out for a pleasant walk. In August 1969, when the photographs above and below were taken, the remains of the wing walls were still in place at the lower end of the lock. The tops of the lock walls had been cut off, and the lock itself filled in during the abandonment. The lock tender's house was still occupied in 1969, at that time by the family of a dealer in scrap metal, and wrecked cars and other iron scrap sat atop and adjacent to the buried lock.

The lock tender's house was abandoned in April 1970 and stood empty until vagrants torched it that August. Over the coming years efforts were made, especially by a group of young people in Dover (Volunteers for Earth, Inc.), to keep the watered canal strip in Hugh Force Park clear, but the stone walls of the house steadily crumbled away as a result of natural conditions and more vandalism. Whatever was left of the lock itself remained buried.

Local interest in the park grew through the 1970s, and that interest was sharpened by the national bicentennial in 1976. That September, Wharton held a Morris Canal Day at the park, and a plaque commemorating the Morris Canal was unveiled near the east entrance to the watered canal strip. Clayton Smith, president of the Canal Society of New Jersey, was the guest speaker at the celebration on September 18. Wharton now holds a Canal Day at the park each summer, and all of this interesting area is now part of the Morris Canal Greenway.

In 2006 Wharton obtained a grant to do exploratory archaeological work at Lock 2 East to determine the condition of the buried remains. The long-term hope was to restore what still existed of the lock in original form and build onto those remains to create a working canal lock. Jim Lee III was the principal investigator for this phase of the work. Local high school students, under the supervision of Wharton's John C. Manna, who headed the lock restoration project, were invited to participate in some of the shovel tests at the site. (Their work led to the discovery of the deteriorating base of a snubbing post, which was duly cataloged.) In the background of the October 2006 photograph above are the much-reduced remains of the stone lock tender's house—just enough standing to hold some graffiti. (James S. Lee III.)

Another photograph from October 2006 shows the excavated west end of the north wall (the side away from the lock tender's house) of the lock. The stones at left show the flaring out at the end of the lock. The stones lying above the excavation are for the most part capstones that were removed from the lock when the ground was leveled during the 1920s abandonment. (James S. Lee III.)

At right is an October 2006 photograph of the excavated east side of the north wall showing the recess for a miter gate in the foreground and a surviving vertical timber (left of center, beside the vertical scale rod) probably used to secure the overlying horizontal wood sheeting that lined this wall in canal times (see page 151). (James S. Lee III.)

Archaeological work continued for several years after 2006. After additional funds were made available, reconstruction of the lock began. This is a January 25, 2012, view to the west from the lower level. (Joseph J. Macasek, Canal Society of New Jersey.)

This view toward the west was taken on October 19, 2018. The lock has been restored almost to its original appearance, and the possibility of this becoming a working lock now appears achievable. The upper drop gate has yet to be added, and the pond beyond the lock excavated and watered. The remaining wall of the lock tender's house rises forlornly in the background.

This view toward the east, also taken October 19, 2018, shows the interior of the completed lock walls and the replica miter gates at the far end. The north wall, at left, includes a replica of the wood wall shown in Olin Vought's 1904 photograph (page 151).

Another photograph from October 19, 2018, from the east end of the lock toward the west, shows the top of the mechanism at the end of the south miter gate. (It appears that someone already had been fooling with two of the hex nuts on the bolts holding the gate mechanism. The Wharton police were notified and sent their thanks. Some graffiti also had appeared. Security cameras would be helpful at this remote place—the fate of the lock tender's house being a case in point.)

At right is a view farther east, along the towpath, looking toward the lock on the same day, October 19, 2018. If you use your imagination, you may picture the girls with their baskets on this same towpath in 1904 and be able to transport yourself briefly back to canal times.

The story of the Morris Canal Greenway across northern New Jersey is unfinished. The building shown above, west of Waterloo, is the former lock tender's house at Lock 4 West, a place locally known as Guinea Hollow. The structure appears to date from about the mid-19th century and probably was built to house the family of Joseph Bird. Joseph Bird was one of the early lock tenders on the canal here in the Musconetcong Valley; his youngest son, Morris Trimmer Bird, was born in 1845, probably raised in this house, and spent most of his working life as the lock tender here. It is the last surviving remnant of a four-generation Morris Canal family and in danger of becoming a lost landmark. The lock itself, which once led to the mile-long slackwater known as Saxton Lake, lies buried behind the house. This property would be a perfect adjunct to a canoe and rowboat launch—a purpose it served in the early 1920s.

The Morris Canal Greenway now encompasses all six of the counties through which the canal passed. In 2012 the New Jersey Transportation and Planning Authority formed the Morris Canal Greenway Working Group to coordinate ongoing greenway projects across the state as a unified effort. Stabilizing and preserving the lock tender's house at Guinea Hollow comprise part of that effort.

Since this new chapter is to an extent about anniversaries and paying homage to some of the people who advocated for the canal in the past, it should be pointed out that more than a century earlier (1912), the New Jersey State Legislature appointed the Morris Canal Investigating Committee to study and decide what should be done with the economically obsolete canal (see page 114). Two years later the committee published its monumental report, which suggested that much of the Morris Canal as it then existed could be repurposed as a recreational water parkway. At about the same time, Montclair businessman Julian R. Tinkham, who independently had recently formed the Morris Canal Water Parkway Association, advocated for the same goal. His group attracted several thousand members, and his plan had the backing of Cornelius C. Vermeule Sr. (who, a decade later, ironically, was tasked by the state with abandoning the canal). The flame of the water parkway idea shone brightly in 1914 but only briefly. By 1929 the canal abandonment by the state of New Jersey was officially completed. As Joni Mitchell sang in the 1970s:

Don't it always seem to go
That you don't know what you've got till it's gone."

AUTHOR'S NOTE

If you are using this book as a reference, this message is for you. When I was given the opportunity to publish this book 20 years ago, I already had been following the Morris Canal as a serious hobby for about 35 years. I had seen errors in existing books and articles, some of which, because they tended to appear over and over, became pet peeves of mine. Among these (as an example) was the misstatement that the Morris Canal was leased to the Lehigh Valley Railroad Company in 1871 for a period of 99 years. In fact, the canal had been leased "in perpetuity," also defined as a period of 999 years. This lease became moot when most of the canal was turned over to the State of New Jersey in 1923, after which the canal itself was drained and the state-owned part of the canal company's property sold off to individuals and communities. Some of the canal's remains escaped demolition for one reason or another, leaving tantalizing clues for later generations to follow.

In attempting to tell the Morris Canal's story my own way, I also made a number of slipups in earlier printings of this book. All of the known slipups have been corrected. I have done my best to ensure that this new edition of *The Morris Canal: Across New Jersey by Water and Rail* is correct as it stands, but I have learned the hard way never to assume anything. If you are using this book to pursue serious research on the history of the Morris Canal and find something to question, please contact me at PO Box 91, Morris Plains, New Jersey 07950-0091. My aim is to get the story right.

—Robert R. Goller
Morris Plains, New Jersey